A History of Dams

A HISTORY OF DAMS

Norman Smith

PETER DAVIES · LONDON

First published 1971
432 151090 0

Printed in Great Britain by
Richard Clay (The Chaucer Press), Ltd.,
Bungay, Suffolk

Contents

Illustrations

Diagrams

Preface

While such branches of civil engineering as bridge-building, road-construction, tunnelling and the laying down of canals and railway systems have already had their histories written, the story of dams has so far attracted the attention of only a few writers; and even they have presented only bits and pieces of the picture and have by no means filled a perplexing gap in the history of engineering. In this book it has been my aim to rectify this situation as much as possible, and to show in the process that dams have a history just as long, and every bit as interesting, as that of other engineering structures. Not only do dams represent some of the most impressive achievements of engineers over the centuries, but their vital role in supplying water to towns and cities, irrigating dry lands, providing a source of power and controlling floods is more than sufficient to rank dam-building among the most essential aspects of man's attempt to harness, control and improve his environment.

Because this book is a historical work the emphasis has been placed quite deliberately on the early periods in dam-building rather than on recent decades. Thus the story is related in some detail down to the end of the nineteenth century but, after that date, is only a survey.

Although the period covered is extensive—from the earliest times down to the present day—the account is restricted geographically to the countries of the Near and Middle East, parts of Africa, Europe and the Americas. The history of dams in India (with a few exceptions) and the Far East (even though Japan's old dams are numerous) has been excluded. These countries are essentially different topics, worthy of study without question, but outside the theme of this book.

While carrying out my research I was financed by the Science Research Council for the first year and for a further two by the Leverhulme Trust Fund, for whom I was privileged to be the first Leverhulme Research Fellow in the History of Technology at Imperial College. I am most grateful for the generous support that these two bodies afforded me. I would also like to record my thanks to the following: Frances Couch and Diana Winchester, who between them typed an immaculate manuscript; my friend and colleague Denis Smith for preparing the drawings so expertly; Mr G. Millhouse, who produced a fine set of photographs from a mediocre collection of negatives; the staff of the Lyon Playfair Library at Imperial College for their assistance, not to mention their good company; and many other individuals and organisations on both sides of the Atlantic who helped me to locate books, pictures and sometimes the dams themselves.

Without exception people have been extremely generous in granting me permission to reproduce photographs, drawings and extracts, and for this I am very grateful.

Two portions of this book have already appeared in print in substantially the same form. The piece on the Roman dams at Subiaco was published in *Technology and Culture*, Vol. 11, No. 1, while much of the material on Spanish dams was the subject of a lecture delivered to the Newcomen Society in December 1968 and will soon appear in its Transactions. To the editors of these journals, Professor M. Kranzberg and Dr S. B. Hamilton respectively, I would like to express my thanks for allowing me to offer the same material again.

Professor and Mrs A. R. Hall have had a special place in the preparation of this book. It was their guidance, encouragement and efforts which gave me the opportunity to study old dams in the first place and without which my findings would never have been written down subsequently. I hope that what I have produced will go some way towards thanking them.

My wife's contribution to my efforts at every stage and in so many ways is quite impossible to acknowledge adequately.

Norman Smith
Imperial College, London
October 1970

Author's Note

It is frequently impossible to give the dimensions of old dams with any certainty and published figures often differ from one source to another. Wherever possible I have checked dimensions on the structures themselves but otherwise it has been necessary to use judgement in assessing which set of figures to quote. Undoubtedly there is still room for correction here and there.

Dimensions taken from written sources have been given in their original units and in the case of metric dimensions the English equivalent is included. There is a good reason for retaining metric measurements in their original form. A figure of 'about 10 metres' is clearly an estimate; a figure of 'about 32·8 feet' is silly; a figure of 32·8 feet is precise and completely obscures the fact that the metric measurement is approximate.

When referring to the left-hand end or right-hand end of a dam I have assumed that the viewpoint is from a down-stream position confronting the air face. This is logical because it is the way a dam is normally viewed.

I

Antiquity

HISTORIANS, AND ESPECIALLY architectural historians, have written a great deal about the building skills of ancient Egypt. Almost without exception they have failed to mention, perhaps because they have been unaware of its existence, one of the oldest civil engineering structures in the world and the oldest known dam. Its intriguing remains have survived to this day near Helwan, some twenty miles south of Cairo. The Sadd el-Kafara, an Arabic name meaning 'Dam of the Pagans', was discovered in 1885 by the German archaeologist G. Schweinfurth, and he and later experts all agree that it was built, during the Third or Fourth Dynasty, some time between 2950 and 2750 B.C. Any dam more than 4½ thousand years old would be notable, but the fact that the Sadd el-Kafara is also a large structure gives the story of dam-building a splendid beginning.

In order to create a reservoir in the Wadi el-Garawi, a dam with a crest length of 348 feet and a base length of 265 feet was required. It was built straight across the wadi at a suitably narrow point, with a maximum height of 37 feet, above the bed of the valley, and it was immensely thick. Because the central section has been washed away it has been possible to examine its internal structure (see Pl. 1 for a general view of the ruined dam). The upstream and downstream portions of the cross-section consist of separate rubble masonry walls, each one being 78 feet thick at the base and reaching to the full height of the dam. Between them these two walls must have contained some 30,000 cubic yards of masonry, perhaps 40,000 tons in weight, and all this material had to be quarried, transported to the site and put into place. But that was not all. The two rubble masonry walls are separated by a space 118 ft wide at the bottom, running the full length of the dam. As soon as the masonry walls had been built, this space in the centre of the dam was filled with 60,000 tons of gravel from the wadi bed and with stones from the surrounding hills. So the completed Sadd el-Kafara contained something like 100,000 tons of material placed in three distinct structural sections, whose combined thickness was 276 feet at the base and nearly 200 feet at the crest: a massive piece of work by any standards.

In order to provide resistance against erosion and wave action, the sloping water face of the dam was covered with carefully placed limestone blocks, each one roughly dressed, and set in rows of steps about eleven inches high (Pl. 2). Evidently this fitted masonry facing was not intended to act as a watertight skin because there is no evidence that any mortar or cement of any description was ever used in the joints between the blocks. Indeed it is widely held that in ancient Egypt mortar was never used as a binding agent at all, but only as a lubricant for easing very large stones into place or to level the top of one building course before starting the next. To us this seems an amazing oversight, especially in the case of a structure designed expressly to hold water. But there are reasons to bear in mind. The massive and monumental structures so characteristic of ancient Egypt were in no great need of cement because their sheer weight was sufficient to ensure stability, as in the case of the Pyramids, for instance. Moreover in a hot country with a sparse and spasmodic rainfall, sealing the walls and roofs of buildings against the elements was largely unnecessary. Had dam-building and hydraulic construction in general been more widely practised, perhaps the need to cement masonry firmly would have been recognised, but this need did not in fact arise. True, there was a great deal of irrigation carried on in ancient Egypt, but always with rather temporary, earth-and-wood diversion dams and unlined water channels. As for permanent dams to create reservoirs, the Sadd el-Kafara was the only one built in Egypt until modern times.

Although nearly 150 feet of the Sadd el-Kafara have been washed away it is still apparent from sighting along the top of the dam that the crest sloped towards the centre. G. W. Murray attributed this to a general settlement of the structure which, as we have seen, was very loosely built.[1] On the other hand it could have been intentional to ensure that overflow from the reservoir only occurred at the centre of the dam and immediately above the bed of the wadi. Because so much of the structure has been washed away it is now quite impossible to verify this notion, or to know if any sort of spillway ever existed at the lowest point on the dam's crest. G. W. Murray and others[2] believed that there never was but, as we shall see, one cannot be sure.

Of one thing we can be certain, however; the dam was only in use for a very short time. As will emerge time and again throughout the story, dams always act as traps for silt. When water flows into the reservoir created by a dam, the silt and sand and other heavy debris carried with the stream are able to settle. The impetus of water moving in a stream or river will carry silt along, but in the still water behind a dam this sediment is deposited on the bottom. Behind the remains of the Sadd el-Kafara there is no evidence of siltation at all, indicating quite clearly that the reservoir must have had a life of a few years at the very most.

All the evidence suggests that the climate around Helwan was much the same in ancient times as it is today. Thus from an examination of the hydrology of the Wadi el-Garawi region it is possible to calculate the volume of water with which the dam had to cope. The catchment area above the dam is 72 square miles. Over this area no run-off will occur unless each fall of rain exceeds 8 millimetres (0·3 in.) but this is likely more often than not. Records for the first part of this century (1904–44) show that in each of three years out of every four the catchment area is likely to receive a rainfall of more than 10 millimetres (0·4 in.) in one day, and that once in every four years a fall of more than 20 millimetres (0·8 in.) in one day is to be expected. So the ancient Egyptians had every right to expect their dam to collect a useful amount of water following an outbreak of rain. Their hopes, as it turned out, were more than realised.

From the remains of the dam it has been possible to calculate that the reservoir's capacity was a little over 20 million cubic feet. A rainfall of 20 millimetres would have been just enough to fill the empty reservoir. A rainstorm of greater than 20 millimetres or two lesser falls, even if they were some months apart, would certainly have over-topped the dam. The rainfall statistics suggest that overflow must have first occurred after only a few years at the most, and since the lack of sedimentation shows that the dam was only in use for a short time it has been reasonably concluded that overflow was the cause of failure.

The dam was very loosely built, and one can well imagine that a stream of water passing over the crest and pouring down the air face would quickly have cut deep into the rubble masonry walls and the sand-and-gravel core. Contrary to previous claims, this does not prove that there was no spillway. It does suggest, though, that if there was a spillway it was not large enough to cope with the volume of overflow which sooner or later was bound to occur. It should also be remembered that the 37-foot-high dam had other structural deficiencies. The poorly constructed body of the dam had no watertight face, and this undoubtedly allowed water to percolate right through the structure despite its vast thickness. In other words the dam experienced erosion not just from overflow but also from 'through-flow'. Another significant feature which may well have contributed to the dam's insecurity is the absence of a cut-off trench. In order to ensure a dam's stability it is normally necessary to excavate a groove along the foundation line into which the dam is then built. This helps to lock the dam into position and also serves to establish a watertight joint between the structure and the valley floor. Ultimately a cut-off trench could not have saved the Sadd el-Kafara, but it is significant that one was not provided.

The poor standard of workmanship was the root cause of the dam's failure, and some criticism of the engineer has been voiced on this account: for instance that he was in a great hurry to put the dam to work and that the shoddy construction reflects his haste. One wonders if this is fair. No one at the time had attempted to build a dam of such size in Egypt or anywhere else. Having no knowledge of what precisely was entailed, the Egyptian engineer of nearly 5,000 years ago should perhaps be excused for doing what seemed to be reasonable, although in fact it proved inadequate.

For what purpose was the dam built? Most ancient dams were built for irrigation, but this was certainly not so in the case of the Wadi el-Garawi, because there is no evidence that agriculture had ever been practised. The only local industry with which the dam can be associated is quarrying.[3] About two miles east of the dam the remains of old alabaster quarries can still be seen, and lacking any other explanation it seems plausible that the water from the reservoir was intended for their use—to provide drinking water for men and animals and also to assist in the extraction[4] of the stone itself. In view of the size of the reservoir it is evident that the intention was to work the quarries on a grand scale—a reasonable assumption for an era when pharaohs were so obsessed with mighty monuments and tombs.

A curious conclusion has been drawn by G. W. Murray as to the significance of the 'Dam of the Pagans' and its failure. As we noted earlier, no other dams were built in Egypt for thousands of years; in fact not until modern times. The suggestion has been made therefore that the failure 'taught the ancient Egyptians never to attempt anything of the sort again'. The likelihood is that the ancient Egyptians simply never needed to build this sort of structure again. Water for alabaster quarries must have been a relatively uncommon requirement, and Egypt had no need of irrigation dams since she depended on the annual flood of the Nile, a unique system of basin irrigation[5] being developed to fertilise the land. Undoubtedly dams were used to regulate this procedure, but they were neither large nor permanent.

The commonly expressed view that a successful Sadd el-Kafara would have significantly altered the course of Egyptian irrigation is not valid, therefore. Only when agriculture required more irrigation than could be provided by the basin system was dam-building undertaken. Irrigation dams on the Nile date from the nineteenth century, and a second attempt at a dam in the desert was not required until 1947.

Is it fair to accept G. W. Murray's view, with which others have agreed, that 'the Sadd el-Kafara is no landmark in the history of engineering'? Certainly the dam had no influence on the development of dam-building.

It was erected too early and did not last long enough to become known in any country where dam-building was poised to begin its development. Moreover the unusual purpose for which the dam was used sets it aside from the mainstream of dam-building history. At the same time, though, Hellström's claim that it is the oldest dam in the world is valid so far as we know.[6] In this sense the Sadd el-Kafara really is a landmark because it is here that the story of dams begins. It is symptomatic of the importance of dams in the development of civil engineering that the first structure was built nearly 5,000 years ago, and remarkable that its surviving fragments enable us to work out so much about its construction, use and eventual failure.

Even though the Sadd el-Kafara really was the only ancient Egyptian dam, this has not prevented other claims being advanced. The two best-known[7] are due to Herodotus, or rather, to be more accurate, to uncritical readers of the 'father of history'. In Chapter 99 of Book II, Herodotus recounts the following tale:[8]

> The priests said that Men was the first king of Egypt and that it was he who raised the dyke which protects Memphis from the inundations of the Nile. Before his time the river flowed entirely along the sandy range of hills which skirts Egypt on the side of Libya. He, however, by banking up the river at the bend which it forms about a hundred furlongs south of Memphis, laid the ancient channel dry, while he dug a new course for the stream half-way between the two lines of hills. To this day, the elbow which the Nile forms at the point where it is forced aside into the new channel is guarded with the greatest care by the Persians and strengthened every year; for if the river were to burst out at this place, and pour over the mound, there would be danger of Memphis being completely overwhelmed by the flood. Men, the first king, having thus by turning the river, made the tract where it used to run, dry land, proceeded in the first place to build the city now called Memphis, which lies in the narrow part of Egypt; after which he further excavated a lake outside the town, to the north and west, communicating with the river, which was itself the eastern boundary.

Of the handful of previous writers on the history of dams, most[9] have taken this portion of Herodotus to mean that King Men (sometimes Menes) dammed the Nile so as to alter its course, and that he subsequently founded the city of Memphis in the old river-bed. This is simply not true.

In the first place Men or Menes is a figure of great antiquity who is believed to have lived some time between 3500 and, at the latest, 2850 B.C. Herodotus was in Egypt probably about the middle of the fifth century B.C., and the notion that the priests of Memphis at a date so long after the

event could produce accurate accounts of the engineering works of Men,
assuming that there were any, is unlikely, to say the least. Furthermore, on
purely technical grounds it is incredible that around 3000 B.C. Egyptian
civil engineering had developed to the point where a river the size of the
Nile could be dammed even temporarily, let alone permanently, as the
story implies. Moreover if the Sadd el-Kafara is typical of what could be
done at the time (*c*. 3000 B.C.) and bearing in mind the fate of this dam,
which was built in relatively ideal conditions, it is evident that Men's
chances of damming one of the world's biggest rivers were remote in the
extreme.

What Herodotus was told, therefore, can be dismissed as legend. Of
greater interest is what he saw for himself, and that, without doubt, was
nothing more than a levee: an artificial flood-protection wall designed to
remove the 'danger of Memphis being completely overwhelmed by the
flood'. It is significant too that the 'mound' was built at a bend in the river,
exactly the place at which the Nile was likely to break out during periods of
excessive flow. The need for the 'mound' or levee and the possibility of
the ancient Egyptians being able to construct it are both acceptable, but
the idea of a dam across the Nile is not, even assuming that one was
needed, which is itself unlikely.

Professor W. Spiegelberg has written of Herodotus that 'he has faithfully
recorded, not history, but stories'. The legend of Men's dam is such a case.
At the same time, however, Herodotus' accounts of what he saw himself are
of more consequence and often very valuable; but not always. Sometimes
even what he claimed to have seen with his own eyes turns out to be false.
This is the case with the curious story of Lake Moeris.

Herodotus deals with the lake in Chapter 149 of Book II and has this to
say:

> The measure of its [Lake Moeris'] circumference is sixty schoenes, or
> three thousand six hundred furlongs, which is equal to the entire length
> of Egypt along the sea-coast. The lake stretches in its longest direction
> from north to south, and in its deepest parts is of the depth of fifty
> fathoms. It is manifestly an artificial excavation, for nearly in the centre
> there stand two pyramids, rising to the height of fifty fathoms above
> the surface of the water, and extending as far beneath . . . The water of
> the lake does not come out of the ground which is here excessively dry,
> but is introduced by a canal from the Nile. The current sets for six
> months into the lake from the river and for the next six months into the
> river from the lake.

Lake Moeris really did exist, and its surviving traces are the so-called
Birket Qarun, a shallow and salty lake in the Fayum depression, west of
the Nile and some sixty miles south-west of Cairo. Very little of what

Herodotus claims for Lake Moeris was ever true. A good deal of research has been done[10] on it, and modern opinion contradicts the older views at virtually every point. The lake is entirely natural, and not an artificially excavated reservoir. Moreover it could never have been used to store Nile flood water (later to be run back into the river) because its level has always been well below that of the Nile.

The canal which Herodotus mentions was genuine, and this was simply an irrigation channel used to bring water to the fertile eastern side of the Fayum, whence excess water was allowed to drain into the lake.

Any notion that Lake Moeris was contained behind a great dam is false, although such an idea has sometimes been expressed. At most there may have been a small regulating dam of earth or wood in the feeder canal with which to control the Fayum's irrigation supply.

Herodotus' estimate of the size of Lake Moeris is much exaggerated, especially the depth, and in fact it is very probable that he saw nothing of the lake for himself. His report, as Hermann Kees has put it, was 'founded on a misunderstanding of Egyptian dragomans' tales'. So once more Herodotus is deceptive, unintentionally no doubt, and the idea of a large artificial reservoir in the Fayum must be rejected.

Perhaps it is disappointing that, with the notable exception of the Sadd el-Kafara, Egypt has so little to offer until recent times. It should not be thought, however, that hydro-technology was absent from ancient Egypt. In particular ways this was far from the case. Because of the Nile's unique régime, the basin irrigation system was a sophisticated and efficient method of watering one of the regions in which man first took steps towards an urbanised way of life. Yet momentous achievement though this was, it did not involve any technological extravagances. To grow his food in a new sociological environment, primitive man did no more than was either feasible or necessary.

The term Fertile Crescent was coined fifty years ago by James Henry Breasted, to refer to the great sweep of the Middle East where civilisation first began thousands of years ago as man gradually gave up food-hunting in favour of food-gathering. Volumes have been written on how this occurred and why it happened in the Fertile Crescent. Of fundamental importance were rivers. At one end of the Crescent was the valley of the Nile; in the middle flowed the Jordan, Barada and Orontes; while in Mesopotamia the Tigris and Euphrates were the sources of water and therefore of irrigation.

The 'two rivers' of Mesopotamia were harnessed for dam-based irrigation at a very early date. Little or nothing has survived of these pioneer schemes in their original form, but traces are believed to exist as part of the hydro-engineering of later societies.

A general point about the irrigation of Mesopotamia should be made at the outset. The Tigris and Euphrates[11] have a huge delta which begins some way to the north of Baghdad and stretches in a broad plain to the Persian Gulf. The two rivers have been slowly laying down this delta since prehistoric times, and as they have worked their way across both rivers have followed meandering courses. Occasionally, major changes in these courses have occurred along with numerous minor ones, something which is typical of any river flowing across very flat terrain. These changes of course make any examination of the history of Mesopotamian irrigation difficult, because it is not always clear exactly where the rivers flowed at various historical periods. Nor is it always easy to decide whether an ancient channel is an irrigation canal or an old river course.

Even though the physical evidence of Mesopotamia's earliest dams has long since disappeared there is little doubt that they were built. The signs are mostly to be found in documentary records, which from the very earliest times are full of references to irrigation. Unlike the Nile's, the natural floods of the Tigris and Euphrates occur at the wrong time of the year for a basin type of irrigation to be practised, and so the only alternative—perennial irrigation[12]—was resorted to. This technique involves the construction of an irrigating canal network whose water supply is controlled by dams. At the right time of year the control dams are opened up to allow the river water to run on to the land. Such a procedure was the key to the very existence of life in Mesopotamia. Irrigation was the prime concern of the ancient city-states of Sumer and Akkad; its problems were frequently a cause of their going to war; and throughout their written records, especially legal codes and religious myths, irrigation is constantly referred to.

From the pre-Sargonic period some plans of canals have survived and also a tablet illustrating a small reservoir. Another tablet from the time of Ur-Nammu, a ruler of the Third Dynasty of Ur (2140–2030 B.C.), contains references[13] to the wages of women who were employed in making a dam of reeds. The use of reeds is also mentioned in a tablet which attributes a dam to Marduk, a Babylonian god of about 1750 B.C., who can probably be associated with the biblical Nimrod. Certainly the following dam was named after this legendary figure.

South of Samarra and north of Baghdad a dam was built across the Tigris in order to prevent erosion and to alleviate the threat of flooding.[14] The intention was nothing less than the diversion of the river into a new course, and that this was achieved is confirmed by archaeological findings. These show that some time in the second millennium B.C. the Tigris experienced a major shift of its course at the very place where the dam, called Nimrod's dam, was built. The name in fact has survived, but, as we shall see later, the dam itself failed around A.D. 1200, allowing the Tigris to return into its old channel.

It is now impossible to be certain how Nimrod's dam was constructed, but since there is no trace of any masonry at its site it is probable that it was made of earth and wood, and indeed the reference on old tablets to 'reeds' is thought to indicate wooden construction. Because it was intended to divert the whole river no water would have flowed over it, and so there was little possibility of the wood-and-earth structure being immediately washed away.

The greatest king of the First Dynasty of Babylon was Hammurabi (c. 1800 B.C.). He carried out extensive irrigation works during his reign, and within his famous legal code embodied rules and regulations pertaining to every aspect of the operation of dams and canals. For instance Section 53 reads thus: 'If anyone be too lazy to keep his dam in proper condition, and does not keep it so; if then the dam breaks and all the fields are flooded, then shall he in whose dam the break occurred be sold for money and the money shall replace the corn which he has caused to be ruined.'

The penalty, then, for not maintaining dams was severe, and this is typical of the irrigational aspects of Hammurabi's Code as a whole. It reflects the immense importance of dams and canals to the very basis of Babylonian society.

The Mesopotamian dam-building record is vague before 1000 B.C. and details are lacking, even though the role of dams can be judged to have been important. Probably they were never very large and usually of inferior construction. With the exception of Nimrod's dam there is no evidence of a structure built across either of the 'two rivers'. But many were built into the banks to control the flow of water into canals.

During the last millennium B.C. the picture begins to clear. Even as Babylon had received life from dams and canals, so it received its death-stroke from them. In 689 B.C. Babylon was attacked and destroyed by the Assyrian King Sennacherib. It is said that to complete the sack of the city he dammed the Euphrates, built up a large reservoir of water and then destroyed the dam. The resulting flood swept through Babylon to complete its destruction. This appears to be the first time, but not the last, that a dam has figured in an act of war.

That Sennacherib was a skilled builder of dams we have ample evidence[15] in the form of surviving structures. He came to the Assyrian throne in 705 B.C. His capital city, Nineveh, stood on the Khosr river, just above its confluence with the Tigris at Mosul. Because the Tigris itself was at too low a level, Sennacherib, in 703 B.C., looked to the Khosr to supply water to Nineveh. We have the king's own word for it that the waters of the Khosr 'from of old took a low level and none among the Kings my fathers had dammed them as they poured into the Tigris'. Sennacherib therefore took steps to dam the river near Kisiri, about ten miles north of

Nineveh, and brought the supply to the city along a canal which he 'dug with iron pick-axes'.

In 694 B.C. Sennacherib was forced to go further afield for additional supplies and became involved in a more elaborate project. From the mountains fifteen miles north-east of Nineveh he diverted and canalised eighteen small streams into the Khosr in order to augment the river's natural flow. This work of canalisation must itself have been difficult in view of the terrain and the lack of anything but the most simple iron tools. But it also involved the construction of two more dams on the Khosr. These were below the point where the canalised supply from the mountains entered the river, and they fed two canals to carry the water to Nineveh. These two dams still exist in a somewhat incomplete form and have been described by R. C. Thompson and R. W. Hutchinson.[16]

The upper dam was the smaller of the two, and rather less of it has survived. It was made of roughly shaped blocks of limestone, sandstone and conglomerate, all firmly mortared together. When the remains of the dam were examined in 1928 it stood to a height of 4½ feet, but originally it may have been higher. The water face was vertical and the air face stepped in five courses of masonry. The base thickness varied from 8½ feet at one end to 9¼ feet at the other. The dam was not built in a straight line across the river: the first 90 feet were at right angles to the river, but then it turned through 90 degrees and the remaining 250 feet lay parallel to the river, gradually merging into the southern bank. The headworks of the irrigation canal appear to have been on the northern bank judging from various masonry remains which have been found there.

The second dam was built 400 yards further downstream and was a much bigger dam, and more of it has survived. Its total length was 750 feet in a long but irregular curve, the northern end being nearly 600 feet further upstream than the southern end. A general view of the dam's remains is shown in Plate 3. The dam had a maximum height of 9½ feet and its cross-section was not uniform throughout. The two ends of the dam feature a vertical water face and a stepped air face, as in the first dam, but in the central section both faces are stepped. This, together with the fact that different types of masonry were used at various places, suggests that the dam has experienced a certain amount of reconstruction and repair at unknown dates (Pl. 4). Near its southern end the dam has been breached and the River Khosr now flows through the gap. There is nothing left to suggest where the canal intake was located.

Some general features of these two important old dams are noteworthy. They were both intended to discharge the river over their crests. At the upper dam a portion of the river's flow was diverted into the canal while the rest poured over the dam. At the second dam this procedure was repeated. Thus it was imperative that the dams should be made of masonry

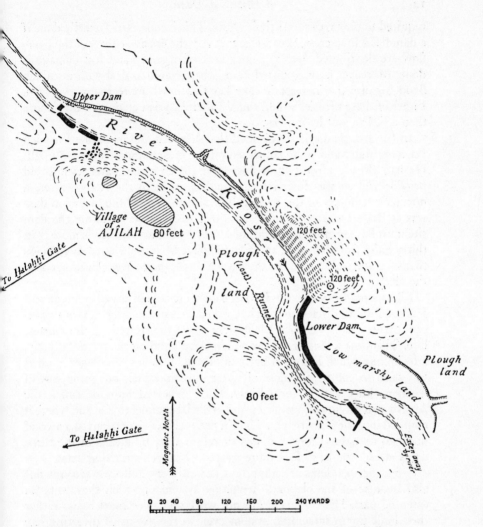

The following labels appear within the figure:

Upper Dam

River

Khosr

Village of AJILAH   80 feet   120 feet

120 feet

Plough (Lost) Runnel land

Lower Dam

Low marshy land

Plough land

To Ḫalaḫḫi Gate

80 feet

To Ḫalaḫḫi Gate

Magnetic North

Eaten away by river

0 20 40   80   120   160   200   240 YARDS

Figure 1   Plan of the River Khosr at the place where Sennacherib built two dams in 694 B.C. (Courtesy of the Society of Antiquaries of London.)

in order to withstand the effects of overflow. It is not altogether clear why the dams are so long, much longer in fact than the river is wide, but two possibilities suggest themselves.

In order to withstand the force of the river, the dams had to be securely fixed to the river-bed. Nature does not provide perfect or even suitable foundations to order, and the curious alignment of the dams perhaps reflects the attempts of Sennacherib's engineers to utilise the best foundation conditions they could find. On the other hand, both dams were required to pass water over their crests. This is more satisfactorily done if a dam has a long crest, because in this way the erosive effects of the over-flow are distributed over a greater area of masonry and less damage is done. Moreover a long-crested dam is better able to deal with a sudden flood because the increase in flow has that much more room to escape. Either of these theories would suggest that Assyrian engineers knew more than a little about hydraulics.

In 690 B.C. Sennacherib decided that Nineveh and its gardens needed yet more water and so he embarked on the most ambitious project of all. He had already tapped the Khosr river for as much water as it could provide and so was forced to go further afield to the Atrush river, a northerly tributary of the Khazir. The site chosen for the diversion dam was at Bavian where the river flows in a deep gorge. Locating the dam site must have entailed some skilful surveying because Bavian is more than thirty miles from Nineveh, and it would have been useless merely to guess that such and such a point on the Atrush was at the correct elevation above the city.

Like its predecessors, the Bavian dam was made of masonry because it was required to discharge the river over its crest. Once again it is signifi-cant that the dam was not built straight across the river along the shortest line. This time a long crest was obtained by building the dam obliquely, a technique which has often been employed. It is not clear what method was used to control the flow of water into the canal, but some sort of wooden sluice gate is most likely. Once deflected into the canal, the Atrush water followed a winding and artificial channel to Kalatah where it emptied into the Khosr river. Thus in 690 B.C. this river received a second boost to its natural flow and was therefore able, by means of the three dams already built, to feed more water to Nineveh's canal system.

Sennacherib's dams are important not only on their own account but also because of the elaborate irrigation system of which they were an essential part. His engineers did not work on a piecemeal basis; rather they built up an integrated system even to the extent of diverting one river into another. For the period the whole project shows considerable sophistication.

Apart from Nineveh, Sennacherib also attended to the water supply of Erbil. This scheme, too, involved combining several different supplies and feeding them into a single river, the Bastara. Some twelve miles north of Erbil, hundreds of blocks of masonry are all that remain of the dam which Sennacherib built across the river to feed the city's canal system.

By the end of the seventh century B.C. the Assyrian empire was in ruins. But the Tigris and Euphrates and their tributaries continued to be crucial to the life of the country and more dam-building was undertaken. In the sixth century B.C. Mesopotamia was absorbed into the Persian empire of the Achaemenians founded by Cyrus the Great in 530 B.C. One of the tributaries of the Tigris on its eastern bank is the Dyala, and on this river Cyrus is supposed to have built an irrigation dam of earth and wood to feed a large network of canals. It was in Achaemenian times too that the first attempts were made at damming the Tigris and Euphrates.

An important feature of the 'twin rivers' is that the Euphrates is higher than the Tigris, and at some period in antiquity this fact was realised, probably in the following way. In Babylonian times the Euphrates followed a much more easterly course than it does today, and this is evident from the sites of places such as Babylon itself, Sippar, Kutha, Kish and Borsippa. Moreover at this time the Euphrates had only one channel, the division into two channels not occurring until some time between 100 B.C. and A.D. 600.

And so we find that for a long period in the second and first millennia B.C. the two rivers in the vicinity of Baghdad were closer than they have ever been since. Regular floods on the higher river caused the banks to burst from time to time, and the excess water then flowed east to the Tigris. In this way flood relief channels were built up between the rivers and these gave the people the clue that was needed to establish an irrigation system. Using the Euphrates as a source and the Tigris as a drain, water was led across country in a primary canal, and from it were supplied a series of interconnected secondary canals. This secondary network fed irrigation water to the land.

Although it is impossible to say exactly when it was first used, the earliest connecting channel seems to have been roughly along a line from Falluja to Baghdad. Originally this canal simply consisted of a part of the Euphrates' prehistoric bed together with a naturally formed flood relief channel. Further to the south other primary canals were created either as entirely artificial connections or as modifications to existing natural channels. In the early stages of their development it is not likely that the primary canals were fed from dams across the Euphrates. Such diversion dams would have been required to pass the bulk of the Euphrates' flow over their crests, and the necessary permanence and structural strength was beyond the engineers of the second millennium. It is likely therefore that small dams or weirs, often referred to as 'side regulators', were built into the river banks at the heads of the canals. This type of structure is adequate for controlling the flow into a canal, but at the same time it is entirely dependent for its supply on the natural level of the river. A river dam is a more sophisticated concept because, quite apart from its ability to divert water into a canal, it is also able to raise the level of the supply to some predetermined height.

Ultimately, however, this deficiency was made good during the middle

centuries of the first millennium B.C. In view of their achievements elsewhere, the likelihood is that the Achaemenian Persians were responsible
for elaborating the irrigation schemes based on the Euphrates and Tigris
and for building dams across both rivers. Such dams were certainly there
in the fourth century B.C. when Alexander the Great was engaged in his
conquest of the Persian empire. Strabo, the Greek geographer of the first
century B.C., for instance, has this to say:[17]

> Now the Persians, wishing on purpose to prevent voyaging up these
> rivers [the Tigris and Euphrates], for fear of attacks from without, had
> constructed artificial cataracts [dams], but Alexander, when he went
> against them, destroyed as many of them as he could, and in particular
> those to Opis.

Whether or not Alexander shared Strabo's view that the dams were
defensive measures is difficult to say. Probably he did not, because we know
that Alexander was keenly aware of the importance of irrigation in
Mesopotamia. Aristobulus, who travelled with him, recounts how
Alexander spent some time inspecting the dams and canals and attending
to a good deal of rebuilding and silt clearance. Moreover he dug a new
canal between the two rivers, called the 'King's Canal', and also reworked
a large artificial channel on the western side of the Euphrates. This, the so-
called Pallacopas canal, was supplied from a dam somewhere near the
modern Hindiya barrage, and at a later date, when the Euphrates eventually shifted its course to the west, it was the line of Alexander's canal
which it chose to follow.

Details of the dams which Alexander found in Mesopotamia and of
those which he added to the system are lacking. It would be foolish to
speculate as to their size or method of construction. But it is relevant to
note that dam-based irrigation was a firmly established practice on the
Tigris and Euphrates as early as the fourth century B.C., and the size of the
canals supplied can be judged from the fact that some of them were
navigable. Even so, the period of the system's maximum development was
yet to come, in the eras of the Sassanian Persians and Abbasid Caliphs.

For all his engineering skill Alexander the Great was above all a general,
and in his work on the Pallacopas canal he was conscious of its potential
use for an attack on the Arabian peninsula. But his death in Babylon in
323 B.C. prevented him from adding Arabia to his already huge empire.
Had he ever marched into Arabia and penetrated to its south-western
corner he would have found there one of the finest of ancient dams and
also the most mysterious. It was built by the Sabaeans.

The Sabaeans were the first Arabian people to develop a civilised form of
society.[18] Their kingdom, Saba, the Sheba of the Bible, corresponded
roughly to modern Yemen, although in their heyday the Sabaean kings

held sway over most of southern Arabia as well. The Sabaeans' rise to prosperity and power was based on trade, by both sea and land routes. They were fine sailors and at an early date (before 1000 B.C.) mastered the difficult job of sailing to and from India and the Persian Gulf and along the treacherous southern coast of Arabia to the mouth of the Red Sea. To Saba these 'Phoenicians of the southern sea' brought merchandise from the Gulf, India, China and Ethiopia. This was then carried north by caravan through Mecca and Petra to Egypt, Syria and Mesopotamia. In addition to the 'foreign' products which they handled the Sabaeans were the main producers in antiquity of spices and incense, especially myrrh and frankincense.

The focal point of the trade routes which the Sabaeans built up was Marib, their capital city. It was in order to irrigate the land around the city that the Sabaeans built the Marib dam, the finest but not the only example of their engineering skill. Marib was built on the northern banks of the Wadi Dhana, a watercourse which is fed by the rainstorms which from time to time inundate the high mountains of western Yemen. It was in order to intercept these floods that a huge dam was constructed across the wadi some three miles upstream from the city (see Fig. 2).

A good deal of nonsense has been written about the Marib dam, and only recently have expeditions to Marib finally ascertained reliable information about the nature of the dam, the way it was used and its history. There have been two principal obstacles to research at Marib. In the first place the site is in a remote and uninviting part of Arabia; and secondly, for a century or more, the rulers of Yemen have steadfastly refused to admit western visitors. A few, such as Edward Glaser[19] in the 1880s, have made the trip in disguise and then narrowly escaped; even a fellow Moslem Arab, Dr Ahmed Fakhry, was only moderately successful.[20] And then in the early 1950s the American Foundation for the Study of Man succeeded in taking an expedition to Marib and they did the best work so far accomplished. What is known about the Marib dam, then, is based very largely on the latter's findings.[21]

Looking at Figure 2, it is a little surprising to find that the dam was not built across the narrowest point in the Wadi Dhana but slightly below it. The probable reason for this will emerge later. The first Marib dam was built around 750 B.C., according to local inscriptions which are difficult to date with any precision. The eighth century B.C. or slightly later does seem very plausible, however, because it was at this time that Saba was becoming a powerful kingdom and Marib an important city. In other words as Marib grew in size, and its population in numbers, the need to develop agriculture emerged. The consequent construction of a dam and irrigation works to harness the only local source of water could only have been undertaken by a powerful ruler with a sizeable labour force at his disposal.

Invariably this is the way in which public works develop. Large urbanised communities create a need to which, in turn, only a large community can provide a solution. And with rare exceptions this can only happen in times of peace. When a state is at war there is no surplus of men, equipment or money to allow the construction of public works. The history of dam-building continually exemplifies this point.

The dam of 750 B.C. or thereabouts was a simple structure of which not a trace has survived. Judging from later developments it was probably made of earth and ran straight across the wadi between the high rocks of the southern side to the rock shelf on the northern bank. Because the dam was made of earth it was imperative that no freak flood should be allowed to overtop it, and so a sizeable spillway was essential. This explains why the dam was not built at the narrowest point on the wadi. By coming 400 feet or so further down the wadi, the engineer was able to leave a gap between the northern end of the dam and the high rocky cliff to the west. It was through this gap that excess water made a northward escape, and here too were situated the intakes of the canal or canals which took water to the city, three miles to the east. The nature of the northern spillway and canal sluices as originally built cannot now be elucidated because they have disappeared under later works.

At the southern end of the first dam it is evident that there were no sluices and no canals. At the time when the first dam was built there was no need to irrigate the southern bank of the wadi, the side remote from the city; and the height of the rock face precluded the provision of a side-spillway at the southern end of the dam.

The earth dam of 750 B.C. was close to 2,000 feet in length and is estimated to have been 4 metres (13·2 ft) high. This estimate is more than a guess even though nothing survives to be measured. It is the minimum height which would have sufficed to ensure a flow of water over the three-mile distance to the gardens which surrounded Marib. Four metres may be a conservative estimate, but nothing less is feasible.

There is little likelihood that the first dam was used to create a reservoir. Its purpose was simply to raise the level of the wadi's flow during periods of run-off, i.e. following a fall of rain in the mountains, and then to divert this water into the canal system. If the notion of creating a reservoir was ever in the mind of the engineer he must have been severely disappointed. The volume of water which the four-metre dam could have stored would have been excessively small. Moreover the idea of a reservoir implies the provision of a low-level sluice in the dam, through which the impounded water could be drawn off. The dam having long since disappeared, it is impossible to know whether a low-level outlet was included; but even if it was, it could only have been used to run water down the wadi bed—too low a level for it to have been much use. There was in addition the problem

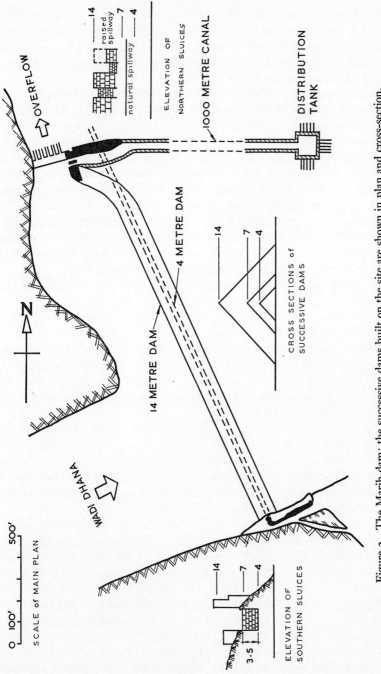

B

N

WADI DHANA

OVERFLOW

SCALE of MAIN PLAN

0   100'   500'

14 METRE DAM

4 METRE DAM

ELEVATION OF SOUTHERN SLUICES

14
7
4

3·5

CROSS SECTIONS of SUCCESSIVE DAMS

14
7
4

1000 METRE CANAL

DISTRIBUTION TANK

ELEVATION OF NORTHERN SLUICES

14   raised spillway
7
4   natural spillway

Figure 2   The Marib dam: the successive dams built on the site are shown in plan and cross-section.
The Figure also shows elevations of the sluices.

of silt. As we noticed earlier when dealing with the Sadd el-Kafara, wadi dams are prone to severe siltation and the Marib dam was no exception. Even today there is clear evidence of thick silt beds above the site of the dam, and these would have seriously reduced the capacity of a reservoir, if such was ever the idea in the first place.

Around 500 B.C., according to more inscriptions, the Marib dam was heightened. Traces of this second structure, 7 metres (23 ft) high, can still be seen, and once again the dam was an earthen bank 2,000 feet long. Its cross-section was triangular with both faces sloping at 45 degrees. The water face was covered in stones set in mortar in order to make the dam watertight and to resist the erosive effects of wave action. Like its pre-decessor, the seven-metre dam was incapable of discharging over its crest, and an adequate spillway was essential. With slight modifications the northern overflow continued to be used. At its northern end the dam swung sharply to the west, thereby confining the overflow to a natural rock surface which was one metre below the dam's crest. It is not clear how the northern canal sluices were arranged in this dam.

The seven-metre dam also supplied water to the southern side of the wadi. To achieve this a channel was cut into the rock face at the southern end of the dam, the bottom of the channel being $3\frac{1}{2}$ metres ($11\frac{1}{2}$ ft) below the crest. When the dam was full this channel, 5 metres ($16\frac{1}{2}$ ft) wide, must have delivered a considerable volume of water. Presumably, though, the outflow could be controlled with some sort of wooden sluice gate, probably a series of wooden planks held at each end in vertical grooves. 'Stop-logs' of this sort have been commonly used throughout history as a simple form of sluice-gate.

The Marib dam was heightened for one simple reason—to increase the area of land which could be irrigated. After 250 years or so Marib had grown in size and population with the result that more cultivation was necessary. The seven-metre dam not only enabled water to be directed to a much greater area of the north bank but also made irrigation of the southern side possible as well.

The next major reconstruction doubled the dam's height and led to elaborate water-works at both ends of the new earth bank, now 14 metres (46 ft) high. At the northern end the floor of the spillway was raised through 5 metres ($16\frac{1}{2}$ ft) and over-flowing water was directed away from the dam along five spillway channels. Irrigation water was drawn off through two masonry sluices in the upstream side of a masonry tank. The space within the tank was used as a settling pond, and this suggests that siltation of the irrigation canals was becoming a problem. From the settling tank the water flowed along a paved channel, 1,000 metres (3,280 ft) long, to a distribution tank from which it ran into fourteen separate canals.

The northern outlet works, whose remains can still be seen (Pl. 5), reached their final form around A.D. 325. The surprising feature is that the overflow is so much lower—3 metres (10 ft)—than the crest of the dam, which suggests that much of the final heightening was worthless. This point is hard to explain. For the fourteen-metre dam the southern sluices were rebuilt and extended as well. To support the end of the earth dam an artificial masonry abutment was built up, and between this wall and rock face ran the outlet canal. This was the same channel that the seven-metre dam supplied, but now its floor was raised $3\frac{1}{2}$ metres ($11\frac{1}{2}$ ft) and masonry walls were constructed to contain the flow. The southern sluices in their final form (Pl. 6) featured masonry construction of very high quality, the carefully cut and fitted blocks using cast lead dowels in their joints but no mortar.

The final form of the Marib dam was reached after the Sabaeans' period of rule in southern Arabia had ended. From 115 B.C. onwards the ruling people were the Himyarites who moved Saba's capital from Marib to Zafar. The fourteen-metre dam at Marib appears to be Himyarite work, and a final attempt to stimulate agricultural prosperity in an area that was already in decline for other reasons. By the fourth century A.D. the rise of Christianity had killed off the incense trade and also led, eventually, to religious wars between Jews and Christians, the latter being directly supported by Abyssinia and indirectly by Byzantium. Even more crippling to southern Arabia was the loss of maritime and caravan trade to the Romans who, from the first century A.D. onwards, proceeded to monopolise commercial traffic between India and the Mediterranean by sailing their ships across the Indian Ocean and up the Red Sea to Egypt. So gradually, during the early Christian centuries, the Sabaeo-Himyarite civilisation decayed and the people migrated. Even when Marib had been a powerful capital city, inscriptions on and near the dam record the numerous repairs which were necessary (quite apart from the two heightenings) when floods broke through the not very substantial earth dam. Once Saba had lost its former ability to organise engineering operations the dam was doomed. It is true that occupying powers such as the Abyssinians made attempts at restoration, notably in A.D. 450 and A.D. 543, but eventually in A.D. 575 it failed for the last time.

A good deal of myth and legend has grown up around this event, including the suggestion in the Koran that God sent a monumental flood to break the dam and punish the people. This tale and others like it are of course attempts to account for southern Arabia's decline in a popular and dramatic way. The real reasons were economic and military—gradual decline over a lengthy period. When the Marib dam ceased to be used, people ceased to maintain it. And so one of the most splendid of ancient dams became a ruin. In passing it is interesting to note how little was known of the dam in antiquity, the only mention occurring in Pliny, who called it the 'Royal Lake'.[22]

Although the Marib dam was easily the biggest one in southern Arabia in antiquity, there were others. Irrigation was extensively practised not only by the Sabaeans but by other peoples at various periods and in different places. From time to time the remains of the dams which were part of this irrigation have been found. Usually they are small dams of earth or masonry or both, and their age is indeterminate. There are a few such structures, pre-Christian in origin, near Aden.[23] At Adraa and Adschma are the remains of two bigger dams which may have been as high as 15–20 metres (50–65 ft) when built.[24] The one at Adraa was particularly well made even to the extent of using twin masonry core-walls within the rubble and earth construction which made up the bulk of the dam. Both sides of it were sealed with stepped masonry facing, which utilised mortared joints and a plastered finish. The Adraa dam was equipped with a masonry spillway, and there is the suggestion of a low-level outlet tunnel. No details of the Adraa and Adschma dams are available, but their relatively advanced design indicates that they are of late date, perhaps even Moslem work.

Another Sabaean dam is known to exist near Marib,[25] and further north in the Najran is a Himyarite dam which features some excellent lime concrete in its masonry construction. Further north still in the Khaibar region (100 miles beyond Medina), six small irrigation dams were built by Jewish communities around A.D. 400.[26] Details of one of them are worth quoting. It was 182 feet long at the base, 270 feet long at the crest and 28 feet high. This dam created a true reservoir because it is fitted with a low-level stone outlet pipe through which the stored water could be drawn off. Unfortunately when the dam was in use it was never possible to use the reservoir's full capacity because its lava banks were so porous.

When the south Arabian civilisation decayed in the fourth and fifth centuries A.D. it is probable that irrigation technology was diffused northwards by the migrating Sabaeans and Himyarties. To some extent this diffusion accounts for the appearance of irrigation dams in central and northern Arabia, such as the Jewish ones near Medina, but another stronger influence was at work as well.

The Nabataeans were a nomadic people of uncertain origin: perhaps they came from Jordan or perhaps from northern Arabia. By the middle of the third century B.C. they had appeared in the Negev, the large wedge of desert between Gaza and Elat which today constitutes the southern half of Israel. Gradually they gained control of it and their kingdom prospered on the profits of the caravan trade bringing goods from Africa and India via southern Arabia—the same trade in fact which had supported the Sabaeans. The 'rose red' city of Petra became their capital.

While the Nabataeans grew prosperous from trade, their other important resource was agriculture. To support a population which was larger
then than it has ever been since, they were obliged to tackle the problem of
cultivating a desert. They were amazingly successful, and not until recent
times has the modern Israeli state attempted the same feat again.

The central Negev around the ancient city of Ovdat exemplifies the
Nabataeans' achievement.[27] The only source of water in this area comes
from occasional showers lasting at most for two weeks per year, and these
may deposit anything between two and fifteen inches of rain—a very low
figure. Moreover the soil has little capacity to absorb water, and the bulk
of the rain produces run-off that lasts for a seasonal total of twenty to
thirty hours at the most. Dams were part of the Nabataeans' technique for
utilising this run-off in various ways.

In the first method dams were built across the beds of wadis in order to
divert the sudden rush of water which followed a rainstorm. The diverted
water was taken by canal directly to the plots under cultivation or else to
stone cisterns where it was stored until required. Alternatively the dams
were used to direct water from one wadi to another if this happened to be
expedient, or to supply short-circuiting canals which carried run-off from
the top of a wadi to the bottom quickly and without loss.

The second technique involved the use of dams to create fertile patches
of ground within their own reservoirs. As noted before, wadi run-off is
liable to carry high concentrations of silt, and the Nabataeans were quick
to learn that this silt is excellent for agriculture. On many wadis, therefore,
the Nabataeans built a series of dams whose purpose was the capture of
water and silt. The whole thing was most skilfully arranged. The crest of
each dam was exactly level with the base of the one above, so that none of
the space between was wasted. The size and location of each dam was so
arranged that the minimum amount of construction work would yield the
maximum area of reservoir.

The creation of these elaborate systems must have entailed some careful
surveying and levelling, but we have no idea how this was done. Nor is it
clear how the Nabataeans estimated just what volume of water and silt
each wadi was likely to yield. Nevertheless these schemes are an early and
splendid example of the use of dams to conserve soil and water, the two
basic elements of agriculture.

When the reservoirs had been filled throughout the length of a wadi,
cultivation could not be undertaken immediately; it took time for the silt
to settle and the water to seep right into it. Then the plots were ready for
planting. The extent to which the Ovdat region was irrigated by these
methods is astonishing; some 17,000 dams have been located in an area of
fifty square miles. None are very large, but their construction is worth a
mention. The dams built to divert water into canals tend to be the largest,

are generally straight and are made of masonry. Their heights, lengths
and thicknesses vary according to individual conditions. A particularly
big one in the Wadi Ovdat is interesting because, in addition to being
fourteen feet thick, it was built to a curved plan, convex upstream.
Probably this was done to meet the prevailing foundation conditions, and
it would not be reasonable to conclude that the dam was designed to
utilise the strength of the arch form in a structural sense.

In the smaller soil- and water-conservation dams the construction is
generally composite—a rubble-and-earth core faced with stepped courses
of roughly shaped masonry blocks. These dams could not have been very
watertight, but then for the purpose involved this was not critical. The
average length of the conservation dams is about 150 feet, their average
thickness is 7 feet and their height about 6 feet. A typical one is shown in
Plate 7.

Taken one at a time these Nabataean dams are in no sense important
civil engineering works. But the system as a whole most certainly is. Faced
with the problem of growing food in a desert whose rainfall was, and still
is, minimal, the Nabataeans worked out a most sophisticated and efficient
solution—indeed, the only solution.

Occasionally evidence has been found[28] of Nabataean dams built to
create small reservoirs of drinking water. One of these is at Rekhemtein
in the great Hismah desert south-east of Aqaba. Across a narrow gorge, a
small masonry dam $5\frac{1}{2}$ metres (18 ft) long, $1\frac{1}{2}$ metres (5 ft) thick and about
1 metre ($3\frac{1}{2}$ ft) high was built to store whatever run-off might follow a
rainstorm. The stored water was used by the local garrison, the remains of
whose outpost have been located near by. Another water-supply dam
occurs at Sela near the modern town of Buseirah where a narrow sloping
cleft in the rocks was dammed at its lower end. Once more the dam is of
large blocks of masonry; its total height was of the order of twelve feet and
its length a little more. At one time a structure of much the same type was
used to store water for Petra.[29]

Whenever there was the slightest opportunity to catch some water, the
Nabataeans built the required dams together with any necessary canals
and cisterns. Their ability both as farmers and traders made them prosperous and powerful, and yet their reign was a short one. Their civilisation flourished only from the first century B.C. to the first century A.D.
when they were overrun by the Roman armies of Trajan. Their irrigation
techniques lived on, however. The Romans were quick to appreciate and
extend the Nabataeans' ideas, and this process was carried on at a later
date by the Byzantines, the latter bringing the Negev to the peak of its
agricultural development.

At this point it is worth while drawing a few conclusions about the dam-
building of antiquity. In the first place it is evident that it was a very basic

technology. Wherever urbanised communities developed in the Middle East, dams were built as an essential part of the need to practise irrigation on which was based the production of food. In passing it is important to note that the same observation is valid for other parts of the world—India and Ceylon, the Far East and pre-Columbian America. When civilisation grew up in these regions dam-building was part of the picture.

The crucial role which irrigation played in the development of the social order in antiquity has been examined in a fascinating study by K. Wittfogel.[30] For the establishment and operation of big irrigation schemes two basic things were required: a large labour force to carry out the engineering work and a strong controlling body, the government, by whom this labour force could be directed. Only in times of relative peace and stability were these conditions likely to be met. The importance of irrigation in antiquity can be judged from the extent to which its operation influenced the machinery of government and society generally. In several cases agricultural areas formed the basic administrative unit. The city-states of Sumeria were single irrigation units and, as mentioned earlier, their wars with each other were frequently caused by squabbles over irrigation problems. In ancient Egypt we find that the hieroglyph for a 'province' is a pictograph of an irrigation system, while the work of various provinces was co-ordinated by a central 'department of irrigation'. In numerous cases a state's taxation system was based on the constituent irrigation units. Hammurabi's Code is but one example of the way in which irrigation influenced the formation and development of legal codes in both Mesopotamia and Egypt. And in technology and science, irrigation was a fundamental stimulus to the development of surveying, hydrology, astronomy and arithmetic.[31] Thus we are dealing with a crucial activity of early society.

Within the early history of irrigation, dam-building shows no overall pattern of development. With rare exceptions such as the Sadd el-Kafara and the Marib dam, dams in antiquity were small structures, and the vast majority were only temporary affairs intended for one season's use. Of the permanent structures which were built, the remains of a few have survived for us to look at and puzzle over. The basic dam-building materials of the ancients were the obvious ones: earth, wood and masonry. The cores of ancient dams were made of earth, sometimes reinforced with wood, or of rubble masonry, or a mixture. Masonry facings were commonly used, and sometimes these were sealed with some sort of mortar. Only in those dams which were required to discharge large amounts of water over their crests did the builders achieve a really solid type of construction, and not always even then.

There is no evidence that the art of dam-building was diffused from one original source. Such a basic technology was fully capable of developing independently at any place where it was needed. At later dates, however, the migrations of people and military conquests probably encouraged ideas to spread. There is not the slightest reason to believe that ancient engineers followed any rules or formulae when building dams.[32] Using whatever materials happened to be available, they built the type of structure which experience told them would be effective. The dimensions of ancient dams were dictated simply by requirements, site conditions and the need for safety, and this last was probably not a critical consideration because very few dams were large enough to be any sort of a threat to life or property.[33]

Dams to create reservoirs were very rare and never large. The idea of storing water during the wet season for use during the dry was not a concept with which the ancients were familiar. Probably the need did not arise. So long as the required amount of irrigation could be obtained from diversion dams across rivers and wadis, people were not likely to experiment with other and more elaborate procedures. In other words, in antiquity, dam-building was widely practised but only in a limited way. It was not until the Romans came on the scene that the size of dams was increased and new uses were found.

2

The Romans

THE ROMANS, WITH every justification, have been called the greatest engineers of ancient times. Their ability to plan engineering works on a grand scale, to carry out the work of constructing them and subsequently to use and maintain their creations was considerable and represents a great advance over their predecessors either in Europe or the Middle East. However, in order to understand Roman engineering properly, it must be remembered that the key to their success depended more on their capacity to organise and administer and on a realistic and practical attitude to technical problems, rather than on any innate flair for engineering as an exercise in itself. Roman engineers did not themselves contribute to engineering very many fundamentally new ideas or original concepts. What they did succeed in doing so well was to absorb all the technical knowledge which came their way and then apply it on a grand scale all over their empire. In the process, engineering underwent a significant degree of development both in its practice and application.

The Romans learnt much of their engineering from the Greeks, who had a marked understanding of technical ideas and were very inventive. Water supplies by means of aqueducts, techniques of tunnelling and harbour-construction are among the engineering activities which were passed on from Greece to Rome. The Greeks' use of Santorin earth to make hydraulic lime appears to have preceded the Romans' own use of pozzolana[1] in mortar and concrete—a development of first importance to Roman engineering and one of the main reasons why so much of their building has survived, and often remained in use, to the present day. It is also clear that the Romans were happy to employ large numbers of Greek engineers and workers on their projects. A basic feature of so much Roman building, and a device which underwent considerable development at the hands of Roman engineers, was the arch. But even this was not a Roman innovation; it was an idea taken over from the Etruscans.

Roman aqueducts, bridges and roads are topics about which a good deal is already known. Through experience and common sense the Romans advanced their design and made significant contributions to the techniques of their construction. A lot less, however, has been written on Roman dam-building, but as we shall now see, they were as competent in this field as in most others.

All over their empire the Romans built large numbers of dams, some of which were very big. Surprisingly, though, so far as is known at present, only three of these dams were in Italy, and all in a group near Subiaco, about fifty miles east of Rome. Our knowledge of them is far from complete, but even so their story is an intriguing one.

In the middle of the first century A.D., Nero established a villa for himself on the banks of the River Anio (the Latin rendering of Aniene) in the gorge above Subiaco. Remains of this villa and other Roman buildings still survive and have been written about from time to time. In order to add to the attractions of his villa, Nero constructed three dams across the river to create three artificial lakes, and for two reasons these dams are of great interest. In the first place they are rare among dams in that they were built for recreational purposes rather than the usual utilitarian ones; and secondly, one of them was the highest dam the Romans built anywhere in their vast empire.

Unfortunately all three structures have long since disappeared, and not much really reliable information about them is available. However they are important enough, especially the biggest one, to merit some discussion. The lakes which the dams created were at different levels, and it seems likely that the upper reaches of the middle and lowest lakes stretched up to the bases of the dams above. In other words the lakes were a series of steps rather like a set of canal locks. It has so far been generally agreed that the first and third dams were small and formed reservoirs of only modest capacity; but the middle dam was large and formed an equivalently large reservoir.

The diagram shows the Aniene valley above Subiaco, the area of the dams. Assuming that the ideas of the nineteenth-century archaeologist John Parker were correct,[2] the positions of the two small dams were as shown in the sketch. But what of the intermediate dam? Its position has never been conclusively identified, but two possibilities have been suggested. The first is directly under the Ponte di San Mauro which carries the road to Guarcino across the river; the second is some 200 yards upstream from the bridge, spanning the gorge between the shell of Nero's villa on one bank and the remains marked 'Roman Arches' on the other.

Of these two possible locations, the latter was the choice of G. Giovannoni.[3] When the road to Jenna was being cut in 1883-4 the remains of
what Giovannoni believed to be the left abutment of the dam were un-
earthed. Giovannoni was able to reach a number of conclusions about the
dam. Its crest, he said, was 13·5 metres (44 ft) thick and paved with tiles.
So that all normal overflow would be constrained to discharge only at the
centre of the dam, the crest sloped towards the middle. Providing access

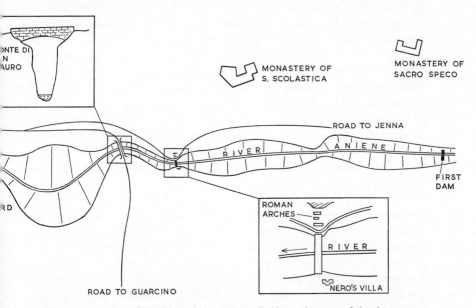

Figure 3   A plan of the River Aniene above Subiaco, the area of the dams.

across the gorge to Nero's villa was a bridge carried along the crest of the
dam on arches for which Giovannoni has recorded dimensions and details
of construction.[4] They were made of concrete with travertine voussoirs
interspaced with large tiles. The piers which supported these arches were
built of massive blocks of travertine and brick-faced concrete.

Was Giovannoni correct in his conclusions? Sir Thomas Ashby at least
was fully satisfied,[5] and there is no doubting the existence of the arches;
they are still there exactly as described. Moreover, as the diagram shows,
at the point where the Jenna road swings round the 'Roman Arches' there
is a distinct promontory jutting out into the valley. Giovannoni claimed
that it was here that the dam abutted against the side of the gorge, and
Ashby agreed. The promontory is a natural formation, and the fact that it
makes the gorge particularly narrow at this point may well have suggested
to Nero's engineers that here was a suitable place for a dam—a very big
one.

Unfortunately the passage of time and the completion of the Jenna road make it impossible to repeat Giovannoni's examination. The tiled crest, for instance, cannot now be seen, and nothing on the opposite bank by the villa survives to provide evidence one way or the other. It is worth remembering, however, that if there ever was a dam on this spot, with a bridge on its crest, as Giovannoni describes, then this bridge must have reached right up to the doorstep, so to speak, of Nero's villa. Giovannoni's argument is at first sight a good one, yet it is not universally accepted. But if he was wrong, what is to be made of the arches and the tiled pavement? Perhaps the arches were merely the vaults of a second and later villa whose remains can still be seen. The tiled pavement may have been a promenade from which to view the river below, and not the crest of a dam at all. Rather than speculating further on this point, however, let us consider the alternative site for the dam.

Under the Ponte di San Mauro was the choice of John Parker, and the site is unquestionably a good one. At this point the gorge is nearly as deep as previously, but very much narrower. A dam as high as 160 feet could have been built here with a length of perhaps 100 feet at the crest and much less at lower levels.

The crux of Parker's theory is the existence in the river-bed just below the bridge of large lumps of the dam which fell there when the structure collapsed. This is a good point. When the dam failed (as will shortly be discussed) what happened to the considerable volume of material of which it was built? A good deal, presumably, has been carried away over the centuries by builders. Some has been washed away by the river which has considerable force when in flood; and a little remains in the river-bed to puzzle historians. These surviving fragments in the river may once have been part of a dam under the Ponte di San Mauro; but they might just as easily have belonged to a dam on the site favoured by Giovannoni. It is simply a matter of deciding how far they have been pushed by the torrent of the Aniene over a period of several centuries. The evidence of the river-bed fragments, then, is dubious.

It is interesting to note, however, that according to Marion Blake[6] there is a great block of concrete among these fragments. Presumably, then, the Romans used concrete for the core of the dam, the finished structure probably being faced with masonry which, as we shall see, was the usual practice. A concrete core would have been a very sound proposition in such a big dam, and this appears to be the first time that a Roman dam featured concrete in its construction.

The strength of the Ponte di San Mauro location is simply that it is an eminently good site for a dam. The Romans were, when all is said and done, good civil engineers. Would they have chosen to dam a wide emplacement, as Giovannoni believed, when a narrow one, under the San

Mauro bridge, was available? Before attempting to answer this question some information on the history of the dam needs to be set out.

In A.D. 38 Caligula began the construction of two new aqueducts to Rome, which were completed by Claudius in A.D. 50 and named 'Claudia' and 'Anio Novus'. Originally the Anio Novus drew its water straight from the River Anio, and then, because this produced a muddy and discoloured supply, a settling reservoir was built but was not entirely successful. We have Frontinus' evidence that during the reign of Trajan modifications to the Anio Novus' intake were undertaken: Frontinus writes, 'He [Trajan] also recognised the possibility of remedying the defects of New Anio, for he gave orders to stop drawing directly from the river and to take from the lake lying above the Sublacensian Villa of Nero, at the point where the Anio is clearest.'[7]

So Nero's dam now found itself serving a useful purpose, supplying water to Rome, and presumably it continued to fulfil this role so long as the Anio Novus was in commission.

What happened to the dams next is largely a mystery. Medieval documents mention two lakes and then one, implying the disappearance ultimately of the first and third dams, as already mentioned. Documents of the period, the twelfth and thirteenth centuries, constantly use the name *pons marmoreus* when referring to the remaining dam, the big middle one, and this is extremely interesting. It lends support to the ideas of Giovannoni who found traces of a marble parapet during his excavations of the dam and bridge.

In 1305 the dam collapsed. Contemporary records place the blame fairly and squarely on the shoulders of two monks who took it upon themselves to remove stones from the dam, apparently in an attempt to lower the level of the lake which was flooding their fields. Probably the dam was not wiped out at one stroke; some decades were presumably needed for the river to carry away the bulk of the structure, aided no doubt by the activities of local builders over a long period. Eventually nothing was left but the fragments which lie in the river-bed; and, according to Giovannoni, the remains which were uncovered by the Jenna road-builders.

Plate 8 shows a painting which hangs in the sacristy of the monastery of Sacro Speco. The monastery stands close by the site of the dams and the painting illustrates incidents in the life of St Benedict, including the years he spent meditating in a cave in the side of the gorge high above the River Aniene. In the painting, which as far as can be ascertained is the oldest surviving illustration of a dam, St Benedict is shown fishing from the crest of the very dam in which we are interested. Although it is not known who the painter was, the picture is generally believed[8] to have been commissioned by Ludovico d'Aragona, Bishop of Majorca, in the year 1428, by which date the dam's failure was already a historical event of 123 years before. Nevertheless, the picture is valuable in throwing light on the nature of the dam and its position. The painting shows a straight masonry wall with two openings through which pours the overflow from the reservoir. These openings presumably represent the arches of the bridge which the Romans built along the top of the dam. At the right-hand end of the structure and immediately adjacent to it stands Nero's villa. Moreover the painting indicates that the top of the bridge was at a level just below that of the villa.

Thus we find that the painter agrees in principle, if not in detail, with Giovannoni; i.e. on the layout of the structure, its position on the river and its height relative to Nero's villa. Almost certainly the painter's dam is short of arches; while he could not have known how many there had been on those parts of the dam which had collapsed, it is disturbing that he does not even indicate arches near the left-hand end—the ones which Giovannoni mentions. Two other points must be considered. Even though the dam had been in ruins for 123 years, surely the painter must have known where it once stood? Enough must have been visible in 1428 to leave him in no doubt as to the dam's location, if not the details of its construction. Secondly, if the dam was in fact at the position claimed by Parker, where the Ponte di San Mauro now stands, then Nero's villa should not appear in the painting right at the end of the dam. Indeed, from the Ponte di San Mauro site the villa is not only a long way off; it is not even visible. In conclusion then, while it cannot be positively proved that the Aniene gorge was dammed in the way Giovannoni described, the evidence is in his favour. There remains, however, the problem of explaining Parker's theory.

One opinion has been withheld from the story so far: Giovannoni believed that there could have been a dam under the bridge as well, but not the big dam that formed the central lake. Rather he thought it was the last dam in the series, the third, and that it was not very large. In the sketch it will be seen that the third lake, according to Parker, was roughly circular in plan, and he believed that this basin had been deliberately excavated by Nero's engineers. Even today, despite thick vegetation, it is evident that at some time the valley has been opened out in the area between the bridge and Parker's supposed site for the third dam. The likelihood is that this excavation was not done merely to create a pool but rather to provide a source of building material, i.e. a quarry. The three dams between them must have required a considerable volume of stone, and the Romans would not have carried all this masonry very far if a local supply was available. Hence a stone quarry close to the dams is an eminently reasonable notion. A huge amount of excavation would probably not have been undertaken just to make a lake, but it would have been unavoidable in order to open a quarry.

A further point substantiates the view that the dam under the Ponte di

San Mauro was the third in the series. As mentioned above, the gorge at this point is very narrow, and the required dam could have been constructed with comparative ease. Nor would it have needed to be very high, because the required lake would only have had to reach a short distance upstream to the foot of Giovannoni's dam. In other words, if it is assumed that the big middle dam was located where Giovannoni suggests, then the site under the bridge is the most logical one for the lowest dam. And below it was the quarry.

From the available evidence, the layout which has now been described seems the most likely one. The location of the first dam, well upstream, has never been disputed. The evidence of Ludovico d'Aragona's painting, the medieval references to a *pons marmoreus*, the excavations of Giovannoni and the nature of the site itself, all support the likelihood that there was a big central dam close to Nero's villa. Below it and under the Ponte di San Mauro stood the third dam, not very high and set deep down in the gorge. Then comes the circular excavation used as a quarry, and beyond this there was no dam at all, despite Parker's claim. In any case the evidence for a third dam below the excavated area is not very substantial. It depends on the interpretation of local nomenclature, while Parker's own opinion is confused and relies to some extent on the notion that the third lake was used to supply the Aqua Claudia. In fact the Aqua Claudia was spring-fed and not connected to the river at all.

Perhaps in the future more explicit documents or some conclusive illustrations will come to light. In the meantime it would be instructive if some excavations were carried out at all the suggested dam sites with a view to discovering where the dams once stood, how big they were and of what they were made. For the time being, however, one final question must be considered. Of the three dams at Subiaco, the middle one was the highest the Romans ever built. But exactly how high was it? The answer is that we do not know and perhaps never will for certain. Giovannoni quotes a figure of 'perhaps forty metres' for the overall height of dam and bridge combined. This figure, however, does not accord with the rest of his theory. If, as Giovannoni thought and the painting appears to confirm, the top of the bridge was level with the present Jenna road at one end and the base of Nero's villa at the other, then the total height of the structure must have been greater. Judging from the writer's own examination of the gorge the figure should be nearer 50 metres (164 ft). But this is still no help in fixing the height of the dam because one does not know how high the bridge was.

Parker, in putting forward his idea of a big dam under the Ponte di San Mauro, raises an interesting possibility by referring to the level of the Anio Novus' water channel. From this it should be possible to work out how high the dam was, because we know that near the end of the first century

A.D. the aqueduct's intake was moved to the lake. Unfortunately, however, the intake has long since disappeared together with an unknown length of the channel. In tracing the course of the aqueduct the first signs of it are in fact near the Ponte di San Mauro. Precise levelling from this point, then, could be used to determine a minimum figure for the height of the dam but nothing more, because it is impossible to work out from the existing remains of the water channel how much higher the intake was. We can be sure of one thing: the Anio Novus' remains are high above the Aniene's present bed, and the crest of the dam could not have been any lower than this point on the aqueduct. In short it seems unlikely that the dam was less than 130 feet high. At the same time we can confidently rule out Lanciani's claim[9] that all three dams at Subiaco were 200 feet high. Such heights would have been not only unnecessary but topographically impossible. Even so, if it is accepted that the biggest of the three dams at Subiaco was 130 feet or so high, it was easily the highest dam which had been built anywhere in the world up to that date. Nothing comparable was to be attempted until 1,500 years later.

It should be added that the position of the dam cannot be fixed by the remains of the aqueduct. The first signs of the aqueduct are consistent with a dam at either the Ponte di San Mauro site or Giovannoni's. But we must remember also what Frontinus said: 'to take [water] from the lake lying above the Sublacensian villa of Nero'. If anything, Frontinus is supporting Giovannoni at this point.

If a big dam were to be built today at the Subiaco site, it would without doubt be an arch dam; the site is ideal for such a structure. While there is nothing to suggest that the Roman dam at Subiaco was arched, there is some evidence that they were familiar with the use of the arch form in dams. This would not be surprising. Having picked up the idea of the masonry arch from their predecessors in Italy, the Etruscans, the Romans made extensive use of it in bridges, both for aqueducts and roads, and also in buildings. In bridges the Romans invariably used the semicircular arch and on occasions built spans of some size. The famous Pont du Gard has lower spans of 74 feet; in the ruined bridge at Narni the main span was slightly over 100 feet; and the single arch of the Ponte San Martino is 103 feet. The Romans' ultimate achievement in the use of the arch form in buildings is to be found in the Pantheon, an incredible piece of work for its time (early second century A.D.) and featuring a hemispherical domed roof 144 feet in diameter.

Such skilful and devoted builders of arches were likely then to try this structural form in dams. In fact the curious thing is that they did not take the idea further; there is only one example known of a Roman dam in which arch action was apparently tried. A second example has been suggested, but the claim is not valid as we shall see.

The town of Glanum, based on an earlier Greek site, was a very early Roman settlement in Narbonensis, an area which corresponds to southern France. Glanum's importance was based on commerce, and it stood at the intersection of two trade routes where its remains can still be seen a mile or so south of St-Rémy-de-Provence. As a Roman town Glanum goes back to the second century B.C. and it continued to flourish well into the Christian era. At precisely what date it was furnished with a water supply is not certain, but this was a standard procedure in most Roman towns. Indeed it is in their towns and cities in France, in places such as Nîmes, Arles and Lyons, that some of the best surviving examples of Roman water-supply systems are to be found. The aqueduct which brought water to Glanum was fed from a small curved dam.

Unfortunately nothing of this structure has survived. In 1891 a new dam was built exactly on the remains of the old one, and this has obviously obliterated all trace of the Roman dam. The nineteenth-century dam is 13 metres ($42\frac{1}{2}$ ft) high, 22 metres (72 ft) long and sharply curved to a radius of 21·75 metres (71 ft). It is significant that in order to obscure the Roman dam the modern one has such marked curvature, but at the same time it is not a true arch dam because it is much too thick at all levels. It is important to emphasise at this point precisely what is meant by three terms: 'arch dam', 'arched dam' and 'gravity dam'. A gravity dam is in general a straight wall of masonry or earth which resists the applied water-pressure because of its sheer weight. In other words the pressure of the water is transferred to the foundations under the dam by means of vertical compressive forces and horizontal shearing forces. The strength of a gravity dam depends ultimately on its weight and on the strength of its base.

In an arch dam the forces due to the water-pressure are carried principally along horizontal lines to the sides of the structure. At the sides the predominantly horizontal forces are resisted by normal forces and shear forces. Underneath an arch dam the only vertical forces are those required to support its weight.

The gravity dam can, for convenience, be regarded as a long series of heavy vertical structural elements of trapezoidal shape firmly anchored at the base. The arch dam on the other hand behaves as a pile of horizontal arches firmly anchored at their ends. It should be added, however, that for actual dams the above views are over-simplified. Gravity dams are not a disconnected series of elements, and arch dams are not a pile of separate rings. Both are monolithic and fixed, so that they are watertight, along the base and up both sides, not just at the points indicated by structural requirements. These factors complicate dam behaviour; nevertheless the simplified models described are still useful.

The third term, 'arched dam', is taken to mean any dam which in plan view is curved, generally to a circular shape but not necessarily so, but which at the same time resists the water-pressure primarily by its weight. In short, then, an arched dam is a curved gravity dam. When an arched dam becomes so slender that it can no longer act as a gravity dam it must be regarded as an arch dam.

In antiquity the vast majority of dams were gravity dams, a few were arched dams, and no example of an arch dam is known. It must also be re-emphasised that the use of an arched dam did not necessarily have anything to do with structural behaviour. More probably curvature was dictated by foundation conditions or hydraulic considerations as previously discussed. These factors have continued to influence the construction of arched dams right down to the present day.

It was an arched dam that was built in 1891 on the site of the old Glanum dam. Although the new dam has completely obscured the Roman one, it is nevertheless still possible to work out something about it. To do this one must resort to the writings of people who were at the site before 1891. There are two sources[10] of information: Esprit Calvet who examined the remains in the 1760s, and Hector Nicolas who was at the site in 1885 but was not, it seems, aware of the earlier work of Calvet.

One of the items examined by Calvet was the aqueduct which took water from the dam to the town. From his measurements it is evident that the dam was of the order of 6 metres (20 ft) high, because nothing less would have been sufficient to raise water to the level of the aqueduct's inlet. All that remained of the dam itself for Calvet to record were the abutments, but what he saw leaves little doubt that the dam when built was markedly curved. At the very least, then, it was an arched dam. The remains of the two ends of the dam also revealed details of its cross-section. It was made of two masonry walls, each a little over 1 metre ($3\frac{1}{4}$ ft) thick, with a space between about $1\frac{1}{2}$ metres (5 ft) wide. Presumably this space was originally packed with earth and stones, while the dam's total thickness must have been something over $3\frac{1}{2}$ metres or about 12 feet.

In other words the Roman dam at Glanum was some 20 feet high, 12 feet thick and curved. On this evidence it does seem very possible that an arch dam was intended. For the time, probably the first century B.C. or a little later, these dimensions of approximately 20 by 12 feet are not at all characteristic of dams intended to resist the water-pressure by weight. Moreover the shape of the river valley at the place where the dam was built supports the hypothesis. It is deep and narrow and, judging from Calvet's drawing,[11] the crest length of the dam could only have been about 30 feet. Its mean radius might have been as much as 40 feet but not more. Such dimensions as can be worked out, then, suggest an arch dam, particularly if the shape of the valley is taken into account.

It was Hector Nicolas, a geologist, who established that the dam was a success. Above the structure he found evidence of recent siltation, suggesting that it was in use for some time before it eventually collapsed. The date of its collapse and much else about the dam remains a mystery: for instance, details of the inlet to the aqueduct, whether or not the dam had a spillway, whether there was a low-level outlet, and what its precise dimensions were. Esprit Calvet never quite got round to an accurate survey; Hector Nicolas did and also took photographs, but neither his drawings or pictures can now be found. The Administration des Ponts et Chaussées, in the process of building the dam of 1891, must have accumulated details of the site, but unfortunately such information as they may possess cannot at present be traced.[12] Should it ever come to light it may very well have something to say about the first arch dam.

The other notable curved Roman dam is in North Africa at Kasserine, about 135 miles south-west of Tunis, and was discovered late in the nineteenth century. This dam seems to have been built to supply irrigation water to the town of Cillium. It is of much later date than the Glanum dam, probably sometime in the second century A.D. The discoverer of the dam, Henri Saladin, has commented[13] on the fine masonry of which it was built, but unfortunately he does not say how it was utilised. However, from his drawings it would seem that cut and fitted masonry blocks with mortared joints were used to face a rubble-and-earth core. Hydraulic mortar was probably required in the construction, and perhaps even concrete. By the second century A.D. both were well established in Roman engineering.

The water face of the dam is vertical, while the air face is stepped through six courses of masonry at the top and vertical below this, a distance of some $12\frac{1}{2}$ feet. The total height is 10 metres (33 ft), the crest thickness 4·9 metres (16 ft) and the base thickness 7·3 metres (24 ft). It has been claimed[14] that a trapezoidal cross-section of these dimensions would not have been used by the Romans in a gravity dam and that in fact they were once more making use of the arch. Certainly the Kasserine dam is not straight. According to one of Saladin's drawings it is markedly curved but not in the form of a circular arc. While the Romans were adept at building segmental arches, it seems highly unlikely that they would have used just any curved shape for a dam, or for anything else. Moreover the dam is very long, about 150 metres (500 ft) according to Saladin, and since it was only 10 metres high it falls far short of the configuration required to produce arch action.

In this case it is much more likely that the dam's curvature was a result of hydraulic and foundation requirements. Saladin's drawings seem to indicate that the topography of the site favoured a curved structure. It is also clear that the dam was required to pass overflow over its crest, the top of the dam being faced with stones set in mortar to prevent scour. Hence it was reasonable to build a curved, and therefore longer, dam to provide more room for the overflow—the typical arrangement.

Like their predecessors, the Romans did not understand the terms 'gravity dam' and 'arch dam' in any formal way. They merely sensed intuitively that sheer weight was one source of structural strength in a dam or, in the single case of Glanum, that arch action was the answer. In the Kasserine dam they provided no arch action and very nearly too little weight.

In his report of 1886 Saladin noted that half the structure was missing, and this has prompted the view [15] that the dam either failed when put into use or perhaps was never even finished, the engineer realising during construction that the scheme was unsound. Neither of these notions is really acceptable, especially the latter. Early engineers had no way of checking their ideas other than a full-scale test of the finished structure. Once the size of a structure had been selected, on the basis of intuition and experience, there was no technical basis on which an engineer could decide to abandon the work, although war, lack of funds or administrative changes sometimes altered the course of events.

Nor is it likely that the Kasserine dam failed in use. Provided it was well made and properly keyed to its foundations, it would have been able, though only just, to preform as a gravity dam. So the Romans succeeded once again but without ever knowing the reason for their success. The passage of time, as was usual, accounts for the absence of half of the dam. We do not know when it went out of use, but since then floods, lack of maintenance and 'quarrying' have probably all contributed to its partial destruction.

A final point of interest is that the dam was intended to create a reservoir and, according to Saladin, a very sizeable one. Low down in the dam wall, at the point where it intersects the river, is an outlet tunnel 2 metres ($6\frac{1}{2}$ ft) wide. Through this the contents of the reservoir were drawn off as required, but by what means is unknown.

The Romans built many other dams in North Africa apart from the one at Kasserine. The remains of many of them have been found, and a particularly important group occurs around Leptis Magna.[16] This region appears to have enjoyed its greatest prosperity when under Roman rule in the second and third centuries A.D., and dams for various purposes were an important part of this phase of Tripolitania's history.

A short distance to the south of Leptis Magna in the Wadi Lebda stands a large buttressed dam which was nothing more than a protection dam. The Wadi Lebda in flood carries large amounts of mud and silt which the Romans were anxious to prevent from reaching the city's harbour. The dam diverted all the unwanted run-off into a canal which led to the sea west of the city. This must be a very rare example of a wadi dam being used to waste water rather than to conserve it.

Leptis Magna's water supply was brought by underground aqueduct from the Wadi Caam, twelve miles to the east of the city. The supply was not in fact from wadi flow but instead from a set of springs in the wadi. These springs, however, are at a very low level, and for this reason a large dam was built half a mile below. This in effect produced a spring-fed reservoir to feed the aqueduct. Details of the dam are lacking, but it has been ascertained that its spillway is at the same level as the inlet to the aqueduct. With a full reservoir, water must have flowed into the aqueduct naturally. For a partially full reservoir, the suggestion is that some sort of mechanical lifting device was used; by the second and third centuries A.D. this is more likely than in any earlier Roman period.

The Romans, for obvious reasons, were very keen on spring water for public supplies, but at Leptis Magna they had to protect the spring-fed reservoir, itself in the wadi, from the muddy water which came rushing down after a rainstorm. A huge dam was the solution. A mile above the aqueduct's intake stands the largest dam in the area, and at 900 metres (2,950 ft) it is one of the longest of all Roman dams. In addition to acting as a silt trap for the benefit of the water-supply dam below, it created several acres of agricultural land whose fertility was replenished after each flood. Details are unfortunately not available, but one end of it appears never to have been finished. Stones projecting from the top of the structure indicate that more courses of masonry were to be added.

The same thing occurs in another big dam built across one of the Wadi Caam's tributaries and shown in Plate 9. The 'starter blocks' are clearly visible, as are two buttresses built against the dam's stepped air face in order to add stability. All the masonry dams which have been found follow the same basic layout. A rubble-and-earth core, sometimes consolidated with concrete, is covered with carefully cut and fitted blocks whose joints are sealed with very strong hydraulic lime mortar. The water faces of the dams are specially sealed with *opus signinum*, a type of plaster made up of hydraulic lime mixed with crushed brick or pottery. The air faces of the dams are stepped to a greater or lesser degree and many of them feature buttresses on the downstream side for added strength. In steep and narrow watercourses where good rock foundations were available, this was the 'standard' Roman dam.

For wide and gently sloping wadis the Romans developed a second standard structure, this time an earth dam. The body of the dam was a simple earth bank, long and low. Probably a masonry facing was used to prevent erosion. However, all that remains of any of the earth dams are the spillways. Since water could not be allowed to flow over the crest, a special masonry spillway, suitably lower in level than the crest, was built into the end of each dam. The flow over the spillways was controlled by stop-logs whose fixing slots are often clearly visible.

With a few exceptions, such as Leptis Magna's water-supply dam, all the Roman dams of North Africa were built for three main purposes: flood-control, water-retention and soil-conservation. As soon as sufficient silt had accumulated in a dam it was used for agriculture, the required irrigation being achieved by wadi run-off. Frequently the dams became silted up right to their crests and were then no longer able to check floods or even retain sufficient water to irrigate the accumulated soil. Consequently many of the dams show signs of having been heightened and strengthened, sometimes more than once. Occasionally their final heights are over 7 metres (23 ft), which allowed the formation of very large and level plots of farmland. The creation of reservoirs of water to be used elsewhere was not a prime function of these dams, but to some extent, following a burst of wadi run-off, this must have happened. There are signs occasionally that irrigation channels were led from the area above a dam for the very purpose of utilising such water as had collected there. All in all the civil engineering works which were built to support life around Leptis Magna and other similar places are impressive in their conception and wonderful examples of how a system of dams can at one stroke solve a related series of hydraulic and agricultural problems.

Furthermore North Africa's collection of Roman dams brings sharply into focus the breadth of the Roman engineer's concept of dam-building. The basic materials of construction, masonry, earth, mortar and concrete, were selectively used in different conditions. Dams were built to divert water and to store it, frequently for irrigation and occasionally for drinking. The role which dams could play in flood-control and soil-conservation was appreciated, albeit on a small scale, while basic hydraulic devices such as overflow spillways, adjustable sluices and low-level outlets in reservoir dams were all utilised.

By now it will be apparent how much similarity there is between the Roman dam-based agriculture of North Africa and the Nabataean schemes in the Negev. This is probably no coincidence. When the Romans absorbed the Nabataeans early in the second century A.D., they inherited the irrigation systems described earlier and were just as dependent upon them. Indeed in Roman times the Negev's irrigation was extended, and the evidence of Roman work is so plentiful that some writers attribute everything to them. C. S. Jarvis, for instance, when he was governor of Sinai in the 1930s, actually took the trouble to renovate an old Roman dam and irrigation system without realising that the work was itself a Roman renovation of an even earlier Nabataean scheme. At the same time, Jarvis' experiment, of which he published a most entertaining account,[17] is of great interest in showing how effective one small dam and its associated irrigation system could be; and that a 2,000-year-old installation could transform life in the desert for a small community even in the twentieth century.

The Romans, then, learnt a great deal about living in a desert from the Nabataean example, and that the necessary technology should percolate along the North African coast in the second and third centuries A.D. is understandable. Roman dams have also been found to the north of the Negev in various parts of Syria. One of these was discovered by Sir Aurel Stein[18] in 1938 at Qasr Khubbaz, about thirty miles west of Hit on the Euphrates. The remains of it are shown in Plate 10 and clearly it is a further example of the standard masonry dam. It is about 20 feet high with a vertical water face and stepped air face of masonry blocks. No details of its length or cross-sectional dimensions are available. Evidently this dam was built to form a reservoir, now full of silt, near the head of a wadi, and the water—two million gallons of it—was used to supply the local Roman garrison established in the area to protect trade routes. The dam dates from late in the second century A.D. and offers interesting evidence that at its maximum extent the Roman Empire's eastern frontier was so near the Euphrates.

Another dam of precisely the same type stands at Harbaka, forty-five miles south-east of Homs. This structure is shown in Plate 11 and was built in A.D. 132. With a height of 18 metres (59 ft) and a length of 200 metres (656 ft) it is a very large dam and was built to create a reservoir either for irrigation or water supply. Today it is full to the brim with silt and no details of its construction are available.

The most impressive Roman dam in the area was near Homs itself. Some eight miles south-west of the city the Romans dammed the River Orontes to form a huge reservoir which has always been known as the Lake of Homs. We must dispose at once of the idea, widely held at the moment, that the Lake of Homs was an Egyptian achievement of around 1300 B.C. In his *Geography*, Strabo has this to say about the sources of the River Orontes: 'These sources are near Mt Libanus and Paradeisus and the Egyptian fortress situated in the land of the Apameians.'[19]

He believed then that the Orontes had three sources, one of which was near an 'Egyptian fortress'. Such fortresses must have existed. During the periods of two of the pharaohs of the Nineteenth Dynasty, Seti I (1313–1292) and Ramses II (1292–1225), Egyptian armies were active fighting the Hittites in what is now Israel and Syria. Notable among their numerous encounters was the Battle of Qadesh, an engagement which Ramses lost and which was subsequently commemorated in a temple painting at Thebes. The painting suggests that the battle was fought on an island fortress, and the probable site has been located on the Orontes and south of the Lake of Homs. Perhaps it was this 'Egyptian fortress' which Strabo associated with the Orontes, although why he thought it was near the river's source is not quite clear. Conceivably he was referring to some other 'Egyptian fortress'.

In 1922 René Dussaud translated[20] Strabo's 'Egyptian fortress' as 'Egyptian wall' and then proceeded to contrive the theory that the 'Egyptian wall' was nothing less than the dam which formed the Lake of Homs. His thesis is elaborate and ingenious but overlooks some basic points. At the time, about 1300 B.C., the Egyptians had no tradition of dam-building in their own country, yet the size and method of construction of the dam, as we shall see in a moment, was clearly the work of experienced engineers, not beginners. Secondly, the Egyptians' occupation of Syria was a precarious one, and short-lived. The area served mainly as a battleground for frequent clashes with the Hittites who for the most part retained the upper hand. The prevailing conditions can hardly be said to have favoured large public works such as dam-building. Thirdly, neither Dussaud nor his disciples have offered any convincing explanation of what the dam was for, nor have they accounted for its size being totally incompatible with the date 1300 B.C. A fleeting and questionable reference in Strabo is not enough to outweigh these objections.

René Dussaud in discussing the dam remarks: 'To conceive the idea, to choose a site discriminatingly, to lay the work out properly and ensure its execution, it was necessary to provide engineers trained in large hydraulic works.' These are entirely apt observations, and the engineers were certainly Romans.

The dam of the Lake of Homs was nearly 2 kilometres ($1\frac{1}{4}$ miles) long in the shape of a very flat V with the point towards the reservoir. The dam was built on an outcrop of basaltic rock which runs right across the Orontes valley, and the alignment of this outcrop explains the shape of the dam. Actually, before the dam was built, the basalt sill itself appears to have formed a small natural lake which the Romans proceeded to enlarge. When the dam was examined in 1881 by Claude Conder[21] and again in 1923 by Léonce Brossé[22] it was in a bad condition. Over the centuries a good deal of damage had been suffered at numerous points; some of this had been inexpertly repaired, and the structure was leaking badly. In spite of considerable mutilation and damage, however, it is still possible for us to gauge what the dam was like originally.

Not surprisingly it turns out to be a typical Roman dam. The core was made of the usual mass of rubble masonry, lumps of locally quarried basaltic rock, in fact, varying in size from an inch or so across to as much as a foot. The whole mass was bound together with a whitish hydraulic mortar, very hard and strong. The core was faced front and back with blocks of basaltic rock cut to a variable size—7–12 inches square by up to 13 inches long. These blocks were in the form of truncated pyramids and were fitted so that the tapered faces projected into the dam while the square faces were exposed. The facing blocks were bound together with hard whitish mortar. Conder has said that the mortar contained traces of

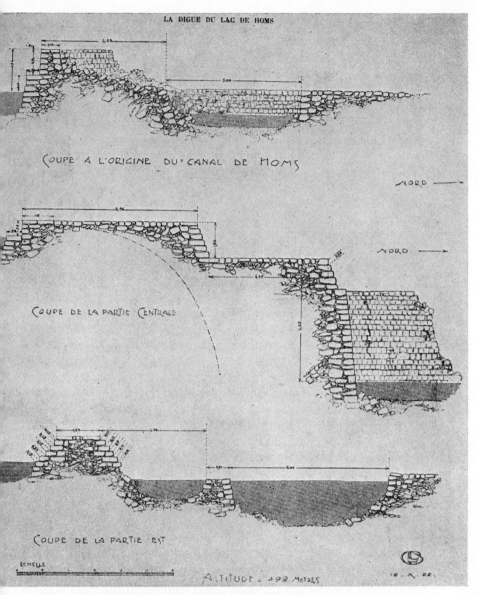

LA DIGUE DU LAC DE HOMS

COUPE À L'ORIGINE DU CANAL DE HOMS

NORD ──

NORD ──

COUPE DE LA PARTIE CENTRALE

COUPE DE LA PARTIE EST

ECHELLE

ALTITUDE = 492 METRES

18 - X - 22.

Figure 4   Cross-sections of the Lake of Homs dam built by the Emperor
Diocletian in A.D. 284. The drawing is that of L. Brossé.

pounded pottery. This may have been used as a filler, in the way that sand
is used, or it may have been added to render the mortar hydraulic—a
device the Romans often used if a natural material such as pozzolana was
not available.

The cross-sectional dimensions of the dam varied along its length. At the eastern end, the crest thickness was about 1·5 metres (5 ft), the base thickness about 3 metres (10 ft) and the height about 1·5 metres (5 ft). Working along the dam to the west, the height gradually increased, as also did the thickness. These dimensions reached their maximum near the centre of the dam and in the western half beyond the point of the V. The greatest height was about 6 metres (20 ft) and the corresponding thicknesses were 7 metres (23 ft) at the crest and perhaps as much as 20 metres (66 ft) at the base. Close to the western abutment the dam's dimensions tapered off. Its water face, as far as we know, was steeply inclined, and the air face was stepped outwards in large increments in the typically Roman fashion.

It is important to note that one cannot be dogmatic about the dam's original size and form. In order to keep it in use for nearly 1,700 years many repairs and modifications were needed. At many points, for instance, it was patched with brickwork to replace the earlier facing blocks which had either fallen off or been stolen. At the western end several large buttresses were added, apparently to discourage the further development of cracks. The dates of these various repairs and additions are not known, but in the end they were not entirely successful. The dam in recent times continued to leak badly despite being well built originally. The suggestion has been made, incidentally, that at some time the structure was perhaps damaged by an earthquake. Originally the Romans faced the crest of the dam with large slabs of rock in order to protect the structure from overflow. But when Brossé examined the dam much of the paved crest had disappeared, revealing the rubble-and-mortar core beneath. In numerous places large sections of the air face steps were missing.

The dam of the Lake of Homs is a late example of Roman engineering, having been built in A.D. 284 by the Emperor Diocletian.[23] He succeeded in creating a reservoir of considerable size—6 miles long by 2½ miles wide. It was the largest artificial lake to have been made up to that time. The water was used to irrigate the fields around Homs and also for water supply. When Conder and Brossé looked at the dam they noted several sluices at various points along the structure, all designed to run water into canals. Only the one near the eastern end seems to have been original. It fed the canal on the eastern bank of the Orontes, the side of the river on which Homs is situated. The other canals are probably later additions designed to take water to the western bank of the river.

Any opportunity to carry out further research at the Lake of Homs dam has now been lost. Plans to construct a bigger dam on top of the Roman one were first mooted in 1881, and after a number of other plans had come to nothing the job was completed in 1934. The builders of the new reservoir can be confident of one thing at least: during nearly 1,700 years of use the Roman reservoir accumulated virtually no silt and in this respect was very unusual.

At the other end of the Roman empire, in Spain, are to be found some of the finest Roman engineering works—bridges, aqueducts and, above all, dams. In 25 B.C. the Emperor Augustus reorganised Hispania into three provinces. One of them, Lusitania, had for its capital the town of Mérida, and in Roman times this place achieved an importance which it subsequently lost and never regained. Unfortunate though this may have been for generations of Meridans, the fact that events passed Mérida by has proved of immense benefit to the historian. The town is unique in Spain for its collection of Roman remains, and some writers even place it beside Rome itself in certain respects. The most splendid monuments include the best specimen of a Roman theatre to be found anywhere in Spain (at present being rebuilt); a small but notably complete amphitheatre; a circus; the remains of an elaborate set of sewers; two road bridges, the one spanning the Guadiana being nearly half a mile long with sixty arches; two aqueduct bridges and two big dams.

The Cornalvo dam is ten miles north-east of the town and was built to collect the water of the River Albarregas. The dam is virtually straight and was very shrewdly placed between two outcrops of rock which make the river valley particularly narrow at this point. It is 20 metres (66 ft) high at the centre and is 200 metres (656 ft) long. The profile of the dam is trapezoidal, the air face slope being 1 in 3 and that on the water face twice as steep. The crest widens, more or less uniformly, from 22 feet at one end to over 40 feet at the other.[24] The maximum base thickness is roughly 200 feet near the centre of the embankment. These dimensions suggest an earth dam, and this, broadly speaking, it proves to be. The point is that a certain amount of masonry construction was also used in a most interesting way. The internal structure of the dam consists of sets of intersecting walls, longitudinal and transverse, which divide the interior of the structure into a series of deep masonry boxes. These were filled either with stones or clay and the whole arrangement was then covered with earth. On the upstream side a masonry facing was provided to stop the water washing the soil away.

It is of course necessary to explain how the nature of the dam's interior has been determined.[25] At the beginning of this century the water face of the dam was in poor condition. Many of the facing blocks had fallen away, and wave action had consequently eaten through the earth covering right down to the masonry core walls. This revealed the general layout of these compartments and the nature of the materials used to fill them, but details of their size, shape and distribution could not be elucidated. Nevertheless the interior construction of the dam says much for the Roman engineers' appreciation of the need for a solid and stable embankment.

When the Cornalvo dam was rebuilt in 1936, the original Roman facing was largely replaced. Formerly the whole of the water face had been covered with large masonry blocks, but these have only been retained in twenty-three sets of steps, each one being the width of a single block and extending about half-way to the foot of the dam. The rest of the face is now covered with small stone blocks laid on to the earth face, with mortar in the lower half but none in the top half, between the steps. The reconstruction also did away with the spillway channel which existed at the left-hand end of the dam. The dam today (Pl. 12) has no high-level outlet.

The low-level outlet, however, is still basically the one which the Romans built. Standing in the reservoir at the foot of the water face is a large masonry tower $14\frac{1}{2}$ feet square and nearly as high as the crest of the dam (Pl. 13). The top of this tower was connected to the crest by means of a masonry arch bridge of which only the springing at the tower end remains. The tower is hollow, and access to the base was provided by means of steps. Originally the tower featured a series of openings at different levels through which water could pass into it. This arrangement was presumably used to ensure that water could be drawn from the reservoir even when silt had accumulated at the base of the tower. It is not known, however, whether the apertures were the only method available for controlling the outflow, or whether there was a valve of some sort in the gallery which runs from the base of the tower through the embankment. Nowadays there is a single large opening near the base of the tower, and the outflow is controlled by means of a modern valve operated from the top.

The water from the Cornalvo reservoir was conducted to Mérida in a covered contour channel most of which has now been lost. It reached the town at a high level on the southern side and is generally believed to have been used for domestic water supply. The history of the dam since Roman times is for the most part a mystery. It seems, however, that it was out of use in the Middle Ages but was once more in commission at the end of the eighteenth century, when a local grandee used the water to drive mills and for irrigation. It is for the latter purpose that the dam is still used.

The other dam at Mérida, called the Proserpina dam, is four miles north of the town (Pl. 14). It collects the water of a small river to form a reservoir which has a perimeter of about three miles when full. At a point where the river valley is suitably narrow, the Romans built a dam 1,400 feet long with a maximum height, near the middle, of 40 feet. The Proserpina dam is built of concrete, masonry and earth, but the layout is not like that at Cornalvo. On the upstream side is a thick wall, running the full length of the dam. Its core is of concrete, and the Romans faced both sides and top with large masonry blocks. The water face of the wall slopes steeply at an angle of 6 degrees to the vertical, while the downstream face is vertical. Unfortunately some of the dimensions of this wall have never been properly determined. It does seem clear, however, that the wall penetrates some 20 feet into the foundations, so that the total height must be of the order of 60 feet. Even the wall's original crest thickness is not certain. Published figures put it at 3·75 metres (12½ ft),[26] but the crest has been damaged, 'quarried' and repaired so much that this figure is difficult to verify. Nevertheless present-day indications are that the top of the wall was in places perhaps only 7–8 feet thick; possibly the Romans varied the

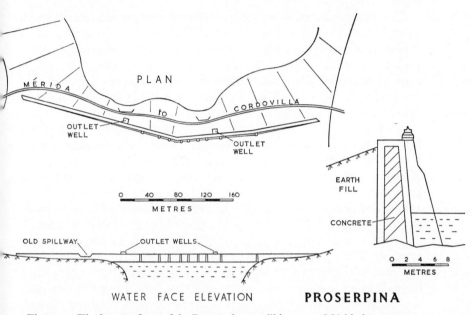

Figure 5   The layout of one of the Roman dams, still in use, at Mérida in western Spain.

thickness from point to point along the dam. Despite a lack of precise data, however, one thing is certain: the concrete and masonry wall does not have a sufficiently heavy profile to act as a gravity dam, and therefore had to be supported at the back in order to resist the weight of the water pressing on the front. The Romans provided the necessary support in the form of a massive earth bank extending nearly 200 feet downstream and sloping more or less uniformly from crest to ground level. The Proserpina dam, then, is a sort of buttress dam in which the buttresses take the unusual form of a continuous earthen bank. This was, however, typical Roman practice which appears in at least one other dam in Spain and in some North African ones also. It is a perfectly viable technique provided that plenty of labour is available to shift and place the soil. There is, however, a built-in danger. When the reservoir is empty the upstream wall is liable to topple into the reservoir under the weight of the embankment. At Proserpina the Romans were well aware of this possibility and countered it by building masonry buttresses on the water face of the wall. Nine buttresses were used. At the top they are some $2\frac{1}{2}$ feet wide and project 3 feet. These dimensions increase with depth, and at the bottom the buttresses project 12 feet. Their spacing is quite arbitrary, and this probably indicates that they were built only where the dam is particularly high.

The Proserpina dam is not straight. Near the centre it swings through an angle of 20 degrees, and then at the left-hand end there is another slight deflection of 5 degrees, as Figure 5 shows. It has been suggested that it was angled towards the water in this way to increase its stability,[27] but this is hardly likely. Such a low and massive structure would not have gained in strength by being 'bent'. The answer undoubtedly is that the most suitable foundation line decided the alignment of the dam wall.

Whereas at Cornalvo the Romans built a single outlet tower separate from the dam, at Proserpina they used two wells built within the main body of the structure. They are situated roughly one-third and two-thirds of the way along the dam. Both wells were constructed in the earth embankment adjacent to the masonry and concrete water-face wall. They are lined with masonry $2\frac{1}{2}$ feet thick, and the openings measure 15 by 12 feet. One is 33 feet deep and the other 50 feet. Both can de descended by means of masonry steps to the level of tunnels which connect with the reservoir. Through these tunnels the Romans controlled the outflow from the reservoir, presumably with sluice gates of some type. Nowadays large-diameter iron pipes with screw valves are used. From the bottom of the outlet wells, the water was discharged through conduits running through the earth bank. It is interesting to note that siltation was a problem in the Proserpina reservoir; and according to Raúl Celestino Gómez[28] it was in the seventeenth century that the outlet tunnels were raised to clear the layers of silt, several metres thick, which had accumulated in the reservoir at the foot of the water-face wall.

A little more is known about the history of the dam of Proserpina than of that at Cornalvo. True, the record is a blank for the Middle Ages, but it appears that in 1617 the governor of Mérida, Felipe de Albornoz, carried out extensive repairs to the masonry covering on the water face, with the result that only on the lower third can Roman work now be found. The same man also capped each buttress with decorative pinnacles, one of which survives. In 1689 a light well was built for one of the outlet tunnels, and then in 1791 repairs were carried out on the smaller outlet well. These minor jobs, of no great interest in themselves, do at least give the impression that the dam has been in use over the past three hundred years or so.

Very extensive repair work was done in 1942, particularly to the masonry wall at the front. The permanent closure of the crest-level overflow, situated near the right-hand end, was one result, and no high-level outlet now exists.

Although the dam is today fully operational and structurally sound, one cannot say it is in a good state of repair. The crest is lacking its masonry covering, exposing very rough concrete and a far from level crest. A good deal of the earth embankment has fallen away, and the remainder is completely overgrown except where a road winds its way from one side to the other. Nowadays the Proserpina reservoir is used for irrigation.

Is it possible to date the dams of Cornalvo and Proserpina? There is no positive information for either, but an estimate can be made. Near the town, the aqueduct from the reservoir of Proserpina crossed the River Albarregas on a bridge, the impressive remains of which still exist. Details of the aqueduct of Los Milagros are not important, but its design is significant. Early bridges for aqueducts such as the Pont du Gard were of the tiered variety, that is, a series of arched bridges built on top of one another. The Los Milagros bridge is of a more advanced design in which one set of piers reach right to the top of the structure, a distance of more than 80 feet. At two levels the piers were braced with arches. Exactly the same design is to be found in the aqueduct bridge at Cherchel in Algeria, an even bigger structure with piers 115 feet high. The Romans used a similar type of construction in both bridges, namely concrete piers faced with stone and brick, and arches of tile or brick with mortared joints. The Cherchel aqueduct is generally attributed to the reign of Hadrian, and it seems reasonable to suppose that developments in aqueduct bridge-building apparent at Cherchel and Mérida date from around A.D. 120–130. It is therefore probable that the dam of Proserpina was built towards the end of the first third of the second century A.D.

The Cornalvo dam is very probably a later work for three reasons. It is, in the first place, further away from the town. As a general rule, the Romans always appear, very logically, to have tapped the sources of water nearest to a town first and gradually moved further afield as the demand increased. Rome is a good example. Secondly, the design of the dam itself is much more elaborate and technically superior. If the Cornalvo dam was built first, one would expect to find the same type of construction at Proserpina. This is not so. The Cornalvo dam represents a structural advance compared with the Proserpina dam, but there is no knowing how long it took for this development to come about. Finally the outlet arrangements at Cornalvo were more elaborate in that steps were taken to prevent siltation from hindering the drawing of water. In other words, if the Romans found that siltation was a problem at Proserpina, then at Cornalvo they took steps to counter it.

The two Mérida dams are the finest surviving Roman achievements in this branch of civil engineering. Although not originally intended for this purpose, they are still able to perform a useful service despite eighteen centuries of use and neglect. Not so fortunate was the dam at Alcantarilla near Sonseca, south of Toledo. Its considerable remains have been described by C. F. Casado.[29] It is precisely the same type of structure as the Proserpina dam. The same type of concrete core-wall is faced with masonry, and in addition there used to be a facing of smaller cut stone blocks on the water face. All but a few of these have been removed by local builders. The upstream wall of the dam is 3·2 metres (10½ ft) thick at the crest, has a vertical water face and a steeply sloping inside face. The dam's greatest height is 20 metres (66 ft) at the centre.

The supporting earth embankment, which ran the full length of the dam on the air face side, once extended some 200 feet downstream near the centre of the dam. But today, both embankment and concrete-masonry wall are breached over a central length of 600 feet out of their total length of 1,800 feet. Unlike the Proserpina dam, this one has no buttresses on its upstream side, and evidently the weight of the earth embankment pushed the concrete-masonry wall into the empty reservoir.

Water was drawn from the reservoir by means of a single outlet well near the centre of the dam. Presumably it functioned exactly as the ones at Proserpina, but being full of earth from the embankment, its interior cannot now be examined. In order to protect the earth embankment from erosion, spillways were built into the dam's crest at each end.

The Alcantarilla dam is of cruder construction than those at Mérida and this is in accord with the view that it was built in the second century B.C. The dam's purpose is clear. It was built to store water for the benefit of Toledo to which city it was connected by means of an aqueduct 38 kilometres long. The Alcantarilla dam appears to be the first ever built to store water for a public supply. The stream across which it stands was not the reservoir's only source of water. Near by the Romans dammed another stream in order to divert a supplementary supply into the Alcantarilla reservoir.

Other Roman dams in Spain—a buttressed dam at Esparragalejo and a few irrigation dams on the River Cubillas near Granada [30]—are small and not worth special mention. What may be the remains of another dam at Lora del Rio have been reported by Thouvenot.[31]

In their use of diversion dams for irrigation and of wadi dams for soil-retention and water-conservation the Romans were copying, on a larger scale, the older techniques of their predecessors. But the application of dams to problems of flood-control and protection appears to be an innovation of their own, and another was the occasional use of dams to regulate canal systems, particularly on Northern Europe's perennial rivers.[32] Their most important contribution, however, was the reservoir dam. To a large extent this was a result of the Romans' concern with water supplies to cities and towns. For this purpose they normally tapped perennial springs and rivers directly, and yet when the need arose, in arid regions such as Mérida, the job of building a reservoir to create a regular and reliable supply was successfully tackled. Works of this sort were made possible by the Romans' ability to plan and organise engineering construction on a grand scale. That they were able to build so many big dams, many of which have lasted such a long time, was also a result of their evolving better methods of construction based on better materials, especially hydraulic mortar and concrete. Moreover they paid proper attention to hydraulic problems, being careful to ensure that water could not percolate through their dams and that, when it overflowed them, spillways were provided to prevent erosion if the crest itself was not suitably protected. All in all, Roman dam-building shows a completeness which was not emulated until the nineteenth century.

C

3

Byzantium and Persia

I T IS CONVENIENT to deal with the story of Byzantine and Persian dams in the same chapter, if only because both were influenced, at the outset at least, by Roman practice. The reigns of the Emperors Trajan, Hadrian and Antoninus Pius, covering in all the period A.D. 98–161, embraced the great period of Roman engineering. By the end of the second century A.D. a decline set in which carried on into the third century. The Romans' former ability to plan public works and to organise their construction and operation was on the wane.

At the end of the third century the real power in the Roman empire began to shift to the eastern half, notably during the reign of Diocletian, builder of the Lake of Homs. Then in A.D. 330 Constantine established the capital of the eastern half of the empire at Constantinople, the ancient Greek town of Byzantium, and subsequently Rome and the western Mediterranean countries of the empire were left a prey to the barbarians. Two types of civil engineering characterise the early history of Constantinople: fortifications and water supply.

The defensive walls of the city were begun under Constantine and were several times enlarged and extended over the centuries. They are an obvious development of Roman techniques, being built with a concrete core and often as much as fourteen feet thick. Frequently a brick facing was used, or sometimes one of large masonry blocks clamped together with iron.

Constantinople's water-supply system goes back to the time of Hadrian, but it is not known for certain what the works consisted of at that period. Probably much of the later work of Valens in the fourth century, Justinian in the sixth century and others was based on Hadrian's original aqueduct scheme. The oldest structure to have survived is the aqueduct bridge of Valens built in A.D. 366 which can still be seen in Istanbul within the area of Constantine's original city. Especially important contributions to Constantinople's water-supply system were made by the Emperor Justinian, who ruled from A.D. 527 until A.D. 565. His was a period in which civil engineering flourished generally: one may mention, for example, the harbours which were constructed in the Roman manner, the numerous fortifications and, above all, the famous church of Santa Sophia in the capital itself.

The very fact that Constantinople was built on a natural defensive position prevented the easy establishment of a water-supply system. Steep slopes ensure rapid run-off of rainfall, and little water seeps into the ground. Moreover summer rainfall is spasmodic and unpredictable. Consequently it was necessary to search elsewhere for a supply of water.

The Golden Horn is fed by two small rivers, the Kiathene Deresi and the Ali Bey Deresi. It was from these rivers, especially the former, that Byzantine engineers took Constantinople's water supply by means of dams.[1] Ultimately eight dams were in use, but for four of these, now out of use, no information has been forthcoming. The other four, however, have not only survived but are still in operation.[2] Three are reservoir dams; one is a diversion dam. The four disused are all diversion dams.

The largest of the four dams, forming the Buyuk Bent or Great Reservoir, stands on a tributary of the Kiathene Deresi near its headwaters, some nine miles due north of Istanbul. The dam is of the gravity type with a total length of 250 feet and a trapezoidal cross-section. The maximum height is 41 feet and the base thickness 33 feet. A rubble masonry core is faced with roughly squared masonry blocks. There are two spillways, one of these being 90 feet long and formed by a long spillway crest at right angles to the end of the dam. These overflow arrangements are not original, however, and it is likely that when first built the dam's own crest discharged excess water. Nor are the present arrangements for drawing water from the reservoir the ones installed in the beginning. The most that can be said of the original system is that it was possible, apparently, to draw water from the reservoir at different levels. Openings in the dam's water face connected with an internal passage that conducted water to the masonry aqueduct leading to Constantinople. How these outlets were controlled is not known.

When full, the Buyuk Bent forms a reservoir a mile or so long, whose capacity is about 500 acre-feet. But in recent years the water level has not been allowed to rise to within ten feet of the dam's crest for fear of a failure. Even when only partially full the dam leaks badly through cracks in its main structure. So although the dam's foundations appear to be sound, the body of the dam is not, in spite of attempts to repair it.

Two miles upstream from the Buyuk Bent is a second reservoir called the Topuz Bendi. The dam is 165 feet long and 23 feet high. Once more the construction is of rubble masonry faced with masonry blocks, and what is essentially a gravity dam is fitted with air-face buttresses. These are very likely a later addition. The volume of water is less than in the first reservoir, and the idea apparently was that the Topuz Bendi should be used to supplement the Buyuk Bent. In other words, water drawn from the higher reservoir through galleries in the dam simply flowed downriver into the lower reservoir when required. This appears to be the first time that two dams were built in series and operated in effect as a single storage unit.

The third dam in the group is a low diversion dam half a mile to the east of the Buyuk Bent. It is intended to raise and divert the whole of the flow of the stream across which it is built into a masonry aqueduct that discharges into the main aqueduct to the city. The most important feature of the dam is its use not only of mortar but also of iron clamps set in lead to bind together the masonry blocks of which the structure is composed. This was a typically Byzantine technique.

The fourth dam of the group is north of the Buyuk Bent and two and a half miles deeper into the Forest of Belgrade. On the Aivat Deresi, a tributary of the Kiathene Deresi, it forms the second largest of the three reservoirs, called the Aivat Bendi. The dam (Pl. 15) is 200 feet long at the crest and is similar in form and construction to the others except that it is not straight; rather it is polygonal in plan, the line of the dam projecting into the reservoir. There is no apparent reason for this irregular shape, and certainly nothing to suggest that it was built this way for structural reasons. Over much of its surface the dam is faced with roughly dressed blocks of marble, used no doubt because marble happened to be available and not because a beautiful finish was desired.

The whole of the crest appears to have been designed to cope with overflow although a three-foot spillway has at some time been added to one end. In common with the dam of the Topuz Bendi, the dam of the Aivat Bendi has recently been equipped with a parapet wall on the water-face edge of its crest. Intended to increase the capacity of the reservoirs, these walls are about three feet high, and the one on the Aivat Bendi dam is supported with a number of small buttresses. The capacity of the Aivat Bendi is less than 250 acre-feet and it feeds a masonry aqueduct which joins the main aqueduct to Constantinople about three and a half miles to the south.

The four dams described are Byzantine but their dates cannot now be determined accurately. One or more of them may be as old as the sixth century when Justinian did much to develop the system. Indeed the fine remains of one of his aqueduct bridges have survived without much change, not far from the sites of the dams. Some of the other aqueducts, however, show clear evidence of later Turkish restoration, and this may apply to the dams as well. Whether or not Turkish rebuilding significantly increased their size is a difficult question. The name of the twelfth-century emperor Andronicus (1183–5) also has been associated with developments to the system.

After a winding journey of several miles from the dams, including crossing many ancient aqueduct bridges, the water was led to numerous reservoirs in Constantinople. Over forty have been identified as Byzantine and the vast majority are underground. Two of the largest were built by Justinian and are still in use. Their ability to store huge amounts of water within the city has led to the suggestion that storage at the dams was not a necessary part of the Byzantines' water-supply system. But it has to be remembered that Constantinople has been besieged many times in its history, and it was this danger that forced the construction of the city's tanks and cisterns. Their existence by no means obviated the need to construct reservoir dams at the supply end of the aqueduct system as well as at the delivery end.

While it is uncertain what Justinian contributed to the system of dams north of his capital, his dam at Daras is a structure for which there is contemporary written evidence. In the third and last of his works, *The Buildings*, written in 560 or soon after, Procopius of Caesarea produced what is essentially a eulogy of his emperor's achievements in public works. Though it may even have been commissioned by Justinian himself, this book is a valuable source of information on his activities as a builder.

In the sixth century Daras was an important town on the frontier with the Persian empire, about eighteen miles north-west of Nusaybin. In 530 Daras was the scene of an important military victory over the Persians, and Procopius gives the place a good deal of coverage. His treatment of the story of the construction of a dam at Daras is for the most part highly imaginative and patently designed to please Justinian: it is not worth repeating; but his description of the finished structure demands quotation.[3]

At a place about forty feet removed from the outer fortifications of the city, between the two cliffs between which the river runs, he constructed a barrier of proper thickness and height. The ends of this he so mortised into each of the two cliffs, that the water of the river could not possibly get by that point, even if it should come down very violently. This structure is called by those skilled in such matters a dam or flood-gate, or whatever else they please. This barrier was not built in a straight line, but was bent into the shape of a crescent, so that the curve, by lying against the current of the river, might be able to offer still more resistance to the force of the stream. And he made sluice-gates in the dam, in both its lower and upper parts, so that when the river suddenly rose in flood, should this happen, it would be forced to collect there and not go on with its full stream, but discharging through the openings only a small volume of the excess accumulation, would always have to abate its force little by little, and the city wall would never suffer damage. For the outflow collects in the space which, as I have said, extends for forty feet between the dam and the outer fortifications, and is under no pressure whatever, but it goes in an orderly fashion into the customary entrances and from there empties into the conduit.

At an earlier point in *The Buildings* Procopius confirms that Daras had suffered damage from floods of the river which flowed near the city walls. Procopius' account is neither as precise nor as complete as one would like, but the essence of the scheme is clear. A dam was built just outside the walls of Daras to contain floods on the river. The two sluice-gates mentioned were apparently at different levels, and it is reasonable to conjecture that one of these was a low-level outlet and the other a high-level spillway. Both would be needed in a flood-control dam: the lower one to draw down the water-level to make sure there was space in which to contain a flood; the higher one to cope with overflow when the dam was full. Once past the dam, the river was apparently led into a channel or 'conduit', and this, to judge from an earlier part of Procopius, carried water into the town.

The most important feature of the dam, of course, was its shape: 'bent into the shape of a crescent', says Procopius. This of itself does not mean that the structure was an arch dam. Its curvature could have been the result of the foundation conditions or the need to provide a long overflow crest. But the dam had a specially built high-level sluice or spillway, and Procopius' additional statement that 'the curve, by lying against the current of the river, might be able to offer still more resistance' must surely be taken to mean an arch dam, the second we have come across. Had Procopius provided dimensions it would be easier to decide whether arch action was utilised here. As it is, we have absolutely no idea how big the Daras dam was, and nothing has survived which can be examined.

What we do know, however, is the name of the engineer, 'a certain Chryses of Alexandria, a skilful master-builder, who [according to Procopius] served the Emperor in his building operations'. Actually Chryses got the job in competition with two other engineers who are much better known. When Justinian first decided to undertake flood-relief works at Daras he called in Anthemios of Tralles and Isidoros of Miletus, famous for their reconstruction of Santa Sophia between 532 and 538. Presumably, then, the Daras dam dates from around this period, but the plan of Anthemios and Isidoros was not the one selected. When Justinian heard of Chryses' alternative scheme he settled for that, although on what grounds is unknown. Certainly we can rule out Procopius' explanation, namely that Justinian had had a vision which concurred precisely with what Chryses had seen in a dream.

Daras was not the only place at which Justinian built a flood-control dam. Procopius says this about Antioch:[4]

> Before that part of the circuit-wall [i.e. Antioch's defensive wall] which happens to be nearest to the ravine out of which the torrent was borne against the fortifications, he built an immense wall or dam, which reached roughly from the hollow bed of the ravine to each of the two mountains, so that the stream should no longer be able to sweep on when it was at full flood, but should collect for a considerable distance back and form a lake there.

This dam also was fitted with sluice-gates, but no other information is available either from Procopius or any other source. Nevertheless it adds to the general picture of dam-building in the sixth century.

Following its conquest by Trajan in the second century, Nabataean civilisation in the Negev experienced a decline in terms of its own prosperity and brilliance. But, as we saw earlier, even under Roman rule irrigation and agriculture continued to be practised according to the Nabataean formula. Under Byzantine rule the Negev experienced something of a renaissance, due in large measure to the reopening of its caravan routes to bring goods to the centre of the Byzantine empire. This renaissance of the fifth and sixth centuries reached its zenith in the time of Justinian. Probably it was then that the population of the Negev was at its greatest and that agriculture was practised on its largest scale. Byzantine dam-building in the Negev flourished most in this period.

In the Wadi Kurnub twenty miles or so south-east of Beersheba are a number of Byzantine dams, perhaps on the sites of earlier Nabataean dams.[5] But whereas any earlier structures were probably small diversion or soil-retention dams, the Byzantine dams were designed to store water for use in times of drought. The fact that there is a chain of dams in the wadi suggests that once more the Byzantines used a set of reservoirs in series as a single storage unit, water from the higher dams being released into the lower ones as required.

The lowest dam in the series and the one in best condition is shown in Plate 16. It was apparently repaired in modern times by the British but nevertheless is now heavily silted, although a certain amount of water can still be stored in it. It is of the standard gravity type with a rubble masonry core and a cut stone air face which slopes steeply. The dam, as far as we can tell, is about 25 feet high and nearly 100 feet long. The deposits of silt have completely obscured details of the crest and water face. Less than 200 feet upstream is another precisely similar dam, also in good condition, usable but not used. The remaining dams higher up the wadi are not now in good condition and appear to have been abandoned at an early date, unlike the two lowest.

The full extent of the Byzantines' dam-building activities in the Negev has yet to be studied. We have far less information than is needed for a full assessment of their contribution to irrigation in the area. The conclusion can be drawn, however, that Byzantine engineers followed in the footsteps of their Roman predecessors. The water-supply system of Constantinople is a clear-cut example of this, while Byzantine irrigation in the Negev represents a logical extension of Romano-Nabataean practice with the addition of storage dams, albeit small ones. The most original piece of Byzantine dam-building is the arch dam at Daras which was built for flood-control.

The story of dams in Persia is one of the longest and most continuous of all. In Chapter 1, mention was made of the dam which Cyrus the Great is supposed to have built across the River Dyala. It was Cyrus who founded the Persian empire of the Achaemenians about 530 B.C. and his successor Darius I who established Persepolis as the capital city. To the south of Persepolis flows the River Kur, and across it the Achaemenians built dams to irrigate the land near the capital. While there is nothing to suggest that any of them have survived in their original form, there is no doubt that some of the ancient dams still standing on the Kur are Achaemenian in origin. One such is the Band-i-Naseri, about thirty miles north-west of Persepolis,[6] and another is the Band-i-Feizabad about twenty miles to the south of the ruins of Darius' city. Of the three dams which he found[7] on the River Kur and claimed for the Achaemenians, the German archaeologist Bergner said one was 25 metres (82 ft) long and 20 metres (66 ft) high. This is surely a mistake; such a large dam was hardly within the compass of engineers of the six or fifth centuries B.C., the period when the Achaemenians were most powerful and when the bulk of their dam-building presumably occurred. To the same centuries we can ascribe the dams in Iraq which, as noted earlier, were found there by Alexander the Great towards the end of the fourth century B.C. On the other hand Sir A. T. Wilson's implication that the Achaemenians also built dams on the River Karun seems to be based[8] on a misunderstanding of Strabo.

It was Alexander the Great who ended Achaemenian rule in Persia, and the dynasties which followed, first the Seleucids and then the Parthians, did not contribute any dams to the engineering scene in Persia. In A.D. 226 Ardashir became the first Sassanian ruler of Persia and the country entered a 400-year-long period of prosperity and stability under a strong central government. Wars were numerous, however, and in the second of his conflicts with the Romans Shapur I succeeded in defeating and capturing the Emperor Valerian along with his army of 70,000 men. This vast body of

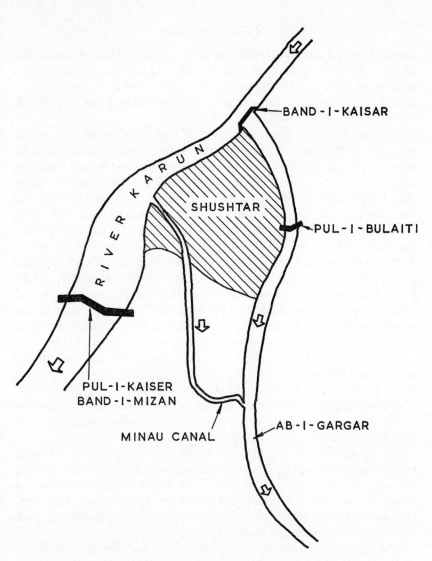

Figure 6 Plan of Shushtar showing the positions of its dams, bridges and canals.

prisoners was dragged off to Shushtar and given the job of damming the River Karun.

Shushtar, built on a high rocky outcrop on the Karun's eastern bank, was an important place in Sassanian times. In order to bring water to the town it was necessary to dam the river to raise its level, and this was first done very early in the Sassanian period. The Minau canal was constructed to conduct water through the town and thence to the high ground to the south. The first dam on the Karun, however, was either too low or too fragile to perform its function satisfactorily, and so the captured Romans were set to work to make improvements. Quite apart from the fact that a complete Roman army constituted a sizeable labour force, it probably included engineers as well.

The first step was the construction of the Ab-i-Gargar (Fig. 6), an obviously artificial cut which still exists.[9] Into it the Romans diverted the whole of the Karun's flow. That this was done is evident from the fact that originally the Ab-i-Gargar carried a very much greater quantity of water than it does now. Once the Karun's bed below the diversion canal had been de-watered, a huge dam was built across the river-bed. With numerous repairs and alterations it has survived to this day and is called the Band-i-Mizan. It is 1,700 feet long and works its way across the Karun in dog-leg fashion, picking up the best foundation line in the process (Pl. 17). It has a rubble masonry core set in hydraulic mortar, and the facing is of large cut masonry blocks held in place with both mortar and iron clamps set in lead. The dam is pierced with numerous sluices for the purpose of releasing water in times of excessive flow. It is not now clear how these were opened and closed, but stop-logs seem very likely.

The Band-i-Mizan only raises the water level a few feet and yet the dam is some 30–40 feet thick. This considerable thickness was provided so that a masonry arch bridge could be accommodated on the dam's crest. The Pul-i-Kaisar or 'Bridge of the Caesar', a reference to Valerian, has survived as well, but not without experiencing considerable damage and rebuilding over the centuries.

The bridge-dam of Shushtar is a notable and typical piece of Roman construction, but quite apart from the care lavished on it, we are told that they also took the trouble, while the Karun was out of the way, to reconstruct the river-bed artificially just above the dam. Perhaps they were anxious to prevent erosion of the river-bed, or, more probably, the work was confined to the river's banks to prevent flooding.

The time taken to build the Band-i-Mizan and the Pul-i-Kaisar is variously reported as from three to seven years. When the work was complete the entrance to the Ab-i-Gargar was closed with a second dam, the Band-i-Kaisar or Caesar's dam.[10] This dam too is still in existence. It is made of large stone blocks mortared and clamped together, and six sluices were provided to control the flow of water into the Ab-i-Gargar. Once past Shushtar, on its eastern side, this canal winds its way south for twenty-five miles and then rejoins the Karun. Ever since the Romans first diverted water down the Ab-i-Gargar, it has been dammed for irrigation, and traces of these structures can still be seen.[11]

The Romans' activities at Shushtar are especially important because for the first time we have evidence of a river being totally re-routed so that a dam could be built in the dry. In wadis or rivers whose flow was very small, for at least part of the year, the job of building a dam's foundations was presumably not difficult; but what was done on rivers with an appreciable flow all the year round remains something of a mystery. When building bridges the Romans certainly resorted to the use of coffer-dams, and perhaps this technique, or some adaptation of it, was employed in dam-building as well. How often ancient engineers went as far as diverting a river completely is impossible to say, probably not very often, because the amount of work involved would have been considerable. At the same time, when dams were built across rivers like the Tigris, the Euphrates or the Karun it is difficult to see how some sort of diversion, even if only partial, could have been avoided.

The Shushtar dams were such a success that other similar projects were undertaken. One of these was at Ahwaz, and until quite recently its remains were still visible.[12] The dam was over 3,000 feet long and, as far as can be judged, about 25 feet thick. For want of any other up-to-date information, the description of the tenth-century Moslem geographer Mukaddasi is worth quoting:[13]

> This [the dam] has been wonderfully constructed from blocks of rock in order to hold back the water behind them. Fountains and wondrous works are there. The dam holds back the water and divides it into three canals which extend to their suburbs and water their seeds. They say that if the dam were not there, Ahwaz could not be cultivated and could make no use of its canals.
>
> In the dam are gates which are opened if the water rises too high; if that were not done, Ahwaz would be drowned. The rushing water creates a roar which makes sleep impossible for most of the year.

Although the Ahwaz dam did not carry a bridge, the one at Dizful on the Ab-i-Diz certainly did, as Plate 18 shows. This dam, 1,250 feet long, is a replica of the one at Shushtar but not quite so old. It was built late in the fourth century A.D. either by Shapur II or by his successor Ardashir II. The structure today is in very bad condition, and following the destruction of a central portion by the river at some unknown date all local irrigation has depended on rough dams of stones and brushwood which have to be reconstructed after every flood.

Yet another bridge-dam was built across the River Karkeh. This dam, or rather its remains, were seen at Pa-i-Pol (Fig. 7) in 1938 by Sir Aurel Stein.[14] It has been out of use ever since it burst in 1837.

One Sassanian dam has survived only in name. Its location, south of Shushtar, where both the Ab-i-Diz and the Ab-i-Gargar flow into the Karun, is still known as Band-i-Kir, meaning the 'Bitumen Dam' (Fig. 7). After Shushtar and Ahwaz this was the most important dam on the Karun,

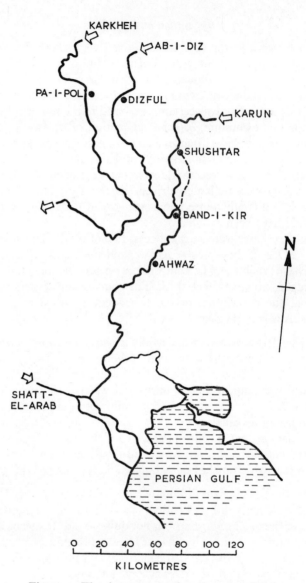

Figure 7   The river systems of south-western Persia.

and the name suggests a rare instance of the use of bitumen to make a dam watertight and solid.

There is no doubt that under Sassanian rule Persia's irrigation experienced a considerable development and was practised on a larger scale than ever before.[15] Nor is there any question that the 70,000 Roman prisoners contributed a great deal to this technical explosion particularly because they would have been experienced in civil engineering construction. And even those dams not actually built by Romans were at least a continuation of the Roman example.

Sassanian dam-building was by no means restricted to the Karun-Karkeh river system of southern Persia. The Sassanians also applied themselves to irrigation works in Iraq, especially on the eastern bank of

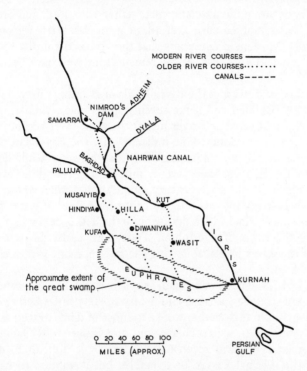

Figure 8   Plan of the Tigris–Euphrates delta.

the Tigris between Samarra and Kut. The complete history of irrigation in this area has yet to be worked out,[16] but the essential points seem to be these. Between Samarra and Baghdad the Tigris is fed by two tributaries on its eastern bank, the Adheim and the Dyala. Both of these rivers were tapped for irrigation water in pre-Sassanian times, and the Dyala in particular was part of an extensive irrigation system, so much so that the river was drained of practically all the water it carried. The Sassanians extended the Dyala's irrigation system and eventually reached the point where they needed more water than the river could offer. The Tigris provided the solution, initially by means of water-lifting devices and then by an enormous canal.

The military successes and administrative powers of Chosroes I (A.D. 531–79) made him one of the most influential of all the Sassanian kings. Irrigation in southern Persia reached the peak of its development in his time, and so it did in Iraq, as is evidenced by the Nahrwan canal. This huge water channel was fed from a dam built across the Tigris north of Samarra, near a place called Dur. At the beginning of this century its remains could still be seen in the river-bed at low water.[17] They consisted of huge stone blocks clamped together with lead. Unfortunately, however, we cannot be sure that these were Sassanian remains. It is known from Moslem geographers that the Abbasid caliphs rebuilt the Nahrwan canal's headworks, so the remains may well have been of their dam.

The Nahrwan canal met the River Dyala at Baquba, thirty-three miles north-east of Baghdad and some seventy miles below the dam. It is of some interest that canal and river met at the same level without weirs or control works of any kind. This indicates that the Sassanian engineers were able to locate their dam in such a way that the water would flow to the level of the Dyala, more than seventy miles away, a levelling exercise of some complexity in the sixth century. Some twenty miles to the south of Baquba the Beldai dam was built to control the flow of water from the Dyala into a short canal which drained into the Tigris below Baghdad and just above the ruins of Ctesiphon. Today this canal forms the lower reaches of the River Dyala, following the failure, many centuries ago, of the Beldai dam.

Below Beldai, the Nahrwan canal, as much as 300 feet wide and 25 feet deep for long stretches, carried on for fifty or sixty miles before draining into the Tigris. On this section were a number of dams to feed secondary canals on both sides, a particularly big one being at Al-Qantara, near Uskaf.[18] A later Abbasid reconstruction makes it difficult to elucidate details of the original Sassanian structure, but it appears to have been about 30 feet high and over 800 feet long in three straight sections, two of which were virtually parallel with the canal and served to guide water over the central section of the dam's crest.

The Sassanian dams and canals on the eastern bank of the Tigris, especially the Nahrwan canal—a massive piece of excavation—were an impressive addition to the region's irrigation. Much about the system is likely to remain obscure, but it is important to note that, just as the Euphrates had earlier been harnessed for irrigation on the Tigris' west bank, so in the sixth century A.D. was the Dyala on its east bank, this river being cleverly combined with the Tigris itself through a series of dams and canals.

Sassanian rule in Iraq was replaced in the seventh century by that of Islam. We shall leave the study of Moslem dams until the next chapter, and in the meantime it will be expedient to look at the post-Islamic period in Iraq and Persia.

Iraq is easily dealt with because there is nothing to report until modern times. In the thirteenth century savage and massive invasions from Asia led to Iraq and Persia falling under Mongol domination. The hordes of Jenghis Khan and his grandson Hulagu have traditionally been held responsible for the devastation of Iraq's irrigation system. This is unjust. As we shall see in the next chapter, the power and authority of the Baghdad caliphs declined drastically in the eleventh and twelfth centuries, with the result that dams and canals were inadequately maintained. Increasingly heavy siltation became an insuperable problem before the Mongols arrived, and equally disastrous was a shift in the course of the Tigris, which by the beginning of the thirteenth century had destroyed the upper sections of the Nahrwan canal. Another significant factor was salination. As the result of several hundred years of irrigation by the successive occupiers of the delta, considerable quantities of salt accumulated on the land. These were deposited mainly by the evaporation and transpiration of irrigation water, and to some extent by capillary action which drew salt from the marine deposits laid down as the delta advanced into the Persian Gulf.

When the Mongols surveyed the agricultural scene in Iraq in the latter half of the thirteenth century, therefore, they saw an irrigation system choked with silt and out of repair, incapable of watering land that was poisoned anyway.[19] So they abandoned Iraq and centred their new empire in Persia. It is here that the evidence of Mongol dam-building is to be found; and it is of considerable significance.

Ghazan Khan ruled Mongol Persia from 1295 to 1304, and his vizier, Rashid-al-Din, has left us valuable details of the Khan's attempts to revitalise agriculture and land-usage.[20] In an effort to bring large areas of Persia under cultivation Ghazan divided agricultural land into three groups, water supply being the governing factor. In the first group were those lands which required very little labour for their irrigation; the second group comprised lands requiring a moderate labour force; while the third group included those where it was necessary to expend a great deal of effort in repairing or constructing a qanat or in building a dam. Rashid-al-Din also records that money was spent on the construction and repair of dams in Khuzistan, a province which he owned himself.

Can any of these Mongol-built dams be located and their characteristics be examined? So far two have been discovered, and both are of great interest. South-west of Tehran, between Hamadan and Qum, flows the Qara Chai river, known in the Middle Ages as the Gavmaha. South-east of Saveh and east of a point where the Gavmaha was joined by two other streams, one from Saveh and one from Aveh, a large dam was built to

provide irrigation water in the dry summer months. According to Mustawfi, a Persian geographer of the fourteenth century, it was built by
Shams ad-Din during the reign of Ahmad, a son of Hulagu and the third
Il-Khan of Persia.[21] The dam can therefore be dated between 1281 and
1284. This one at least, and there may have been others, pre-dates the
dam-building mentioned in the reforms of Ghazan Khan. No contemporary accounts of Shams ad-Din's dam have been found, but fortunately
the structure survived into modern times and details are available.[22]

    The dam (Pl. 19) is always referred to as the Saveh dam even though
that town is not the nearest one to it. Its most striking feature is undoubtedly its size; with a height of 60 feet it is among the highest to appear in
our story so far. The crest length is about 150 feet. In plan the structure is
straight and was designed to resist the water-pressure by its weight; in
other words it is a gravity dam. It is made of rubble masonry without any
facing blocks. The water face is stepped at approximately four-foot
intervals, giving the whole thing a curiously stratified appearance. Running down the water face of the dam from crest to foundation is a peculiar
curved mass of masonry projecting into the reservoir. It is clearly visible in
the photograph and measures some 20 feet wide at the crest and 40 feet at
the bottom. This protrusion is obviously an integral part of the dam and
contains a number of openings at various levels, by far the largest being at
the bottom, about 12 feet above the foundations. This arrangement is new
and will be discussed in more detail below.

    Two other things about the Saveh dam are significant. Firstly, it is
rather surprising that this massive but crudely built structure has survived
for nearly 700 years apparently without suffering damage or deterioration.
Secondly, the water-face pictures clearly indicate that the reservoir has
not silted up to any noticeable degree, which in view of the dam's age
seems to suggest a river remarkably free of silt. In fact both features are
easily accounted for; the dam has never been used. The unfortunate
engineer, whoever he was, chose to build the Saveh dam at quite the
wrong place. The limestone rock at the sides of the valley was sound as far
as is known, but the foundations at the base of the dam were not. The
structure was built on river alluvium consisting of sands and gravel, which
modern investigation has shown to go down 90 feet before bed-rock is
reached. What happened is obvious. As soon as the reservoir began to fill
the pressure above the foundations was easily sufficient to drive the water
through the alluvium, and the water immediately established a permanent outlet for itself. The engineers of the time had no answer to this, and
unhappily we must record that, according to local accounts, when the
engineer saw his mistake he committed suicide. In this his judgement was
perhaps rather better. If he had not taken his own life, it is highly probable
that someone else would have.

It is difficult to estimate the amount of blame which should attach to the engineer of Saveh. It is an easy matter for us in the twentieth century to point out where the fault lay. The need for thorough site investigations is fully recognised today because the failure of a dam is rightly looked upon as an extremely serious occurrence. But what was the situation centuries ago? It is a fact of building history that, in general, only the successes survive. The failures cannot usually be examined, either because the structure was a total loss or because it was rebuilt into a successful form. Perhaps the Saveh dam suffered a not unusual fate for its times, but it is unusual in that it survived even though it could never fulfil its purpose and was abandoned.

Whatever the inadequacies of the foundations at Saveh, in two respects the engineer chose his site well. The narrow gorge was ideally suited to provide a considerable reservoir from a dam not especially large in total volume. This at least reflects an intelligent approach to the problem. Hydrologically the site was a good one too. The river carried a sufficiently large amount of water, particularly in the winter, to ensure a full reservoir for the summer drought. These facts are fully confirmed by a modern plan to resurrect the Saveh dam site (although not the old dam). The idea was first mooted in the 1930s and again after World War II. The plan is for a rock-fill dam 230 feet high, providing a reservoir capacity of 25,000 million gallons. Had our Mongol engineer been successful, even a 60-foot-high dam would have captured a lot of water. But the modern work has proceeded slowly and has been beset by serious problems (in the foundations of course).[23] It is unfortunately too late to console the Saveh engineer of 700 years ago. It should finally be recorded that his work was not an absolute failure; the dam has at least served as a bridge across the river.

The second dam of the Mongol period is at Kebar, about fifteen miles south of the ancient town of Qum. The dam (Pl. 20) was discovered in 1956 by Henri Goblot, a French engineer working in the area, who has published details.[24] It was built in a roughly V-shaped gorge which suddenly narrows, about half-way down, to a deep gully, much deeper than it is wide. The rock is limestone, and this time there were no foundation problems. The dam, which has survived intact, is 26 metres (85 ft) high and 55 metres (180 ft) long at the crest. The crest thickness is at no point less than 4·5 metres (15 ft) and nowhere more than 5 metres (16½ ft). The air face if vertical, except near the base where there is a slight slope in the downstream direction. In other words the dam thickens slightly in the deep gully near the bottom. Much of the water face of the dam is today obscured by the vast amounts of silt and debris which have collected in the reservoir. Where it can be observed at the top, the water face is vertical, and it seems reasonable to conclude that this face of the dam is vertical throughout its height. We have a dam, then, 85 feet high, 180 feet along the crest, and for the most part rather less than 20 feet thick. The Kebar dam is thus a very thin structure, too thin to act as a gravity dam. It is in fact an arch dam, the oldest surviving example of this type of structure so far located.

In addition one further dimension of the Kebar dam should be given: the radius of curvature of its air face is 38 metres (125 ft) at all points, the dam constituting what is known, in modern terminology, as an arch dam of constant radius. The radius of curvature of the water face is more difficult to determine because of the thick beds of silt which obscure much of the face, and even where it is visible the curvature is not well defined.

At Kebar the engineer recognised the need to provide very secure connections between the dam and the sides of the valley. Up both sides of the dam the limestone rock was cut away to form a groove as wide as the dam is thick. The dam is built into these grooves, presumably to their full depth. Not surprisingly the depth of these slots cannot be determined, but it is clearly adequate because the dam's abutments have performed their function satisfactorily. The loads applied by the dam to the rock have not caused any cracking or slipping, while from a hydraulic point of view the joint is still perfectly watertight. The foundation groove is carried right to the base of the dam and underneath it at the bottom of the deep gully where the dam is only a few feet long. Thus the base of the Kebar dam is also fully fixed to provide not only structural resistance but also a watertight joint at the most important point: the water-pressure at the foot of a dam is higher than anywhere else. Today the base of the dam is still completely leak-proof.

The construction of an arch dam at Kebar is highly significant for two reasons. In the first place the builders recognised, quite correctly, that the site was ideal for an arch dam; and secondly, having chosen this type of structure, they adopted a constant radius–constant thickness form, securely keyed into the rock. As a result of a great deal of careful research in the past fifty years, the modern dam engineer could point to many defects in the Kebar dam. Constant dams are not the best; the thickness should vary from top to bottom of an arch dam, while even the circular form is unsuitable in some instances. It would, however, be unrealistic, not to say unreasonable, to apply hindsight to Kebar. Although it is the most elementary kind of arch dam, it represents a notable development in the story. Moreover, it is appreciably more slender than the Roman dam of Glanum appears to have been.

The form of the dam at Kebar dealt with, we can now examine its other features. It has a core of rubble masonry set in mortar. The faces are cleanly finished with roughly dressed rectangular blocks of small and varied size. The facing blocks have mortared joints but are not closely fitted. We know that the builders quarried the limestone locally because both the dam and the local stone are of exactly the same type, and in any case there would be no point in bringing materials from a great distance; yet the quarry, which would not have been very large, has not been found. Presumably over the centuries erosion and rock falls have made it un-recognisable, a common fate of ancient quarries. The mortar which was used is called locally *saroudj*. It was and still is made from lime crushed with the ash of a desert plant which is used in the first place as a fuel. The addition of ash makes the lime hydraulic and results in a strong, hard and highly impervious mortar ideal for dams, and undoubtedly an important factor in the Kebar dam's long life. The use of this material in Mongol Persia, a tradition which apparently goes back a long way, throws an interesting light on the cement which the Abbasids may well have used.

As far as the supply of water to the reservoir was concerned, the engineers chose a good site. The Kebar river flows all the year round and is the only one in the area not contaminated with gypsum salts. It is interesting, however, that the capacity of the reservoir is very small compared with the total annual potential of the river: 404 acre-feet compared with several times that amount in the river. However, the reservoir was presumably big enough to fulfil its purpose.

What exactly was this? It seems clear that the dam was originally built for irrigation. Not far away and connected to it by an open water-channel are a number of tanks or cisterns. In the absence of any evidence of villages or settlements in the area, it is likely that the cisterns were for agricultural and not domestic use. Subsequently, however, the dam was made to serve a different purpose. At the beginning of the seventeenth century, during the period of Safavid rule in Persia, a caravanserai was built beside the cisterns. The remains of this still exist and have been positively identified as Safavid work. Presumably it was built near the dam so as to be able to use the existing water-supply. A significant feature of the dam perhaps supports this conclusion: it has been heightened. About 7 feet from the present crest, and on both faces, there is a distinct join in the masonry, marked on the air face by a step 16 inches wide. That is to say, the top 7 feet is 16 inches thinner than the main body of the dam. Moreover while the main body of the dam is an arch, the seven-foot section at the top is curved only where it joins the original structure. At the ends it changes into two straight walls, 16 feet long at one side of the valley and 40 feet at the other. These end walls are also 7 feet high.

The dam was almost certainly raised in this way in an effort to combat the siltation of the reservoir, which was gradually reducing the water capacity. Henri Goblot estimated that the heightening increased the reservoir capacity by between 120 and 160 acre-feet.[25] It does seem very likely that this attempt to restore the dam's usefulness is connected with the construction of the caravanserai, in which case its heightening dates from the early or middle seventeenth century. There is, however, no proof of this.

We come now to the point where the Saveh dam takes the stage again. On the water face of the Saveh dam was a curved protrusion containing a vertical series of openings, the lowest by far the largest. The Kebar dam is also provided with a set of water-face apertures although the layout is slightly different. The projecting bulge of masonry is absent and the openings are flush with the water face. Moreover at Kebar there are two series of openings instead of one, the series being aligned more or less vertically and some 5–6 feet apart. The apertures in one set are rather bigger than in the other: 3–4 feet square compared with $1\frac{1}{2}$ feet square. Whereas at Saveh all the holes are still visible, at Kebar siltation of the reservoir has obscured all but four, two in each series. The number and position of those below the silt is therefore conjectural and based on the configuration of those which can still be seen. The series of large apertures is in line with the deep gully at the base of the dam, but whether or not the holes extend that far down is not clear. The line of smaller openings is nearer the side of the valley and probably too far across to intersect the gully at all. Inside both dams, the small openings are connected by various shafts and galleries which have yet to be fully explored and explained. It is clear, however, that the tunnels eventually emerge on the air-face side, and this much is certain: the apertures, shafts and galleries were there to provide passage for the water through the dam walls.

This must have been the technique employed to draw water from the two reservoirs. So long as at least one aperture was below the surface of the reservoir, the water could be drawn off. But why were there so many openings? Why not one at or below the lowest level to which the reservoir was likely to drop? Siltation is probably the answer. The builders must have known that, as the reservoirs silted up, a single low-level outlet would soon be blocked. Therefore they provided a series of openings, accepting the fact that in time each one would be successively blocked until only the top one was operational in a reservoir of much-reduced capacity. Whether or not they ever entertained the idea of desilting the whole reservoir in order to recommence the cycle is not known, but this is extremely unlikely.

At Kebar there is only one outlet tunnel on the air-face side. This indicates quite clearly that the tunnels from the various water-face apertures connect inside the dam and then emerge as a single channel on the air face. A very simple arrangement suggests itself. Behind the water-face openings is a vertical shaft into which the water-face holes discharged their contents; at the bottom of this shaft, a horizontal gallery ran through the dam to the air-face side. Sometimes it must have been necessary to stop the discharge from the reservoir altogether. It is not at all clear how this was done. The obvious solution would have been a valve operating in the horizontal outlet tunnel, but the existence of such a thing and the mechanism to control it have yet to be found. It is relevant to note, however, that at Kebar it is still possible to descend into the main body of the dam by means of a large vertical shaft adjacent to the line of small openings. The shaft reaches down about twenty feet and is fitted with a narrow set of spiral stone steps. Whether or not this access shaft has anything to do with opening and closing a valve or valves is not clear. It has been established, however, that the access shaft is connected by openings at the side to the space behind the series of small holes. For the dam at Saveh, about which rather less is known at present, the purpose of the bulge on the water-face side can now be deduced. It contains a vertical shaft into which the reservoir water, but for the foundation failure, would have discharged via the small apertures. From the bottom of this well there must be a tunnel to direct the water to an air-face outlet. At both Saveh and Kebar, once the water emerged from the air-face outlets it was channelled away for use.

So much for the small apertures. What of the big ones? There is one at the base of the Saveh dam and at least two, still visible, at Kebar. It must be said immediately that their purpose is not definitely known; but at least one plausible explanation can be offered. As we have noticed several times now, all dam-builders are faced with the problem of keeping the workings dry during construction. At Kebar and Saveh there was no possibility of using the dry season, because both rivers are perennial; and the use of a temporary diversion dam, the method most frequently used today, was either unattractive or impossible. Instead, large holes were left in the dams through which the river could pass for most of the year at least. Presumably if the discharge capacity of the holes was exceeded, the unfinished dam overflowed and work was abandoned. When the dams were at last finished, the holes were filled permanently. We know that they were not blanked off on the water-face side, because there they are still visible, and this suggests that closure was effected on the air-face side.

In the light of the above idea the use of one large hole at Saveh and several at Kebar is interesting. At first sight the Saveh arrangement appears to be more logical. If the first and lowest opening was big enough to cope with the whole of the river's flow, no more would have been needed higher up. Yet Kebar has them. Henri Goblot's explanation of this is intriguing. He claims that each aperture represents one year's work, and that at the end of each period of construction a new outlet was built and the one below filled in. Hence he concludes that the Kebar dam was built in four or five years because there were probably four or five outlets—the two still visible above the silt and two or three others whose presence is conjectural.

This interesting theory depends in the first place on the assumption that each stage of construction occupied a year or part of a year. That may well be true. It is very likely that, even with a diversion tunnel through the dam, the river flooded the workings during the wet season, and no construction was possible. But M. Goblot's argument does not answer the main objection. At Saveh the engineer used only one outlet, logically enough near the base of the dam. Why therefore was not one also sufficient at Kebar? Why was it necessary to use several at different levels, only one of which was in use at a time? Two possibilities suggest themselves. It is conceivable that Goblot's theory of successive closure of the holes is wrong, and that actually they were all in use until the dam was finished, when they were all filled in at once. This would certainly have improved the efficiency of the system; but it would not explain why the engineer at Saveh managed with one. Alternatively the engineer at Kebar was perhaps in a hurry and, in order to derive the maximum benefit even from an unfinished dam, kept the temporary outlet as high as possible. The position is anything but clear. That the large openings in the dams were used to divert the river during construction is plausible enough, but the exact procedure followed is a mystery. Perhaps more thorough study of the structures will provide an answer.

It is useful to turn next to the overflow problem (actually at Saveh, as things turned out, it was 'underflow' that was important—and disastrous). The Saveh dam has no spillway, nor has that at Kebar. Obviously in periods of overflow it was the intention to release excess water straight over the crests. There is clear evidence that this happened at Kebar. The crest shows traces of debris carried by the water, and at one end the side of the valley has been worn away by the flood current to a height of three or four inches above crest level. In addition the crest of the dam has been damaged at certain points, although this may be due to natural deterioration rather than flood damage.

The Saveh dam has been positively dated from contemporary accounts. It now remains to show that the dam of Kebar is of similar age. Certain characteristics common to both structures are helpful here. They are, in the first place, both high dams designed for water-storage. This approach to the irrigation problem had not, so far as we know, been tried before in Persia, and certainly not on anything like the scale of Kebar and Saveh. Previously, under the Abbasids, irrigation needs were met by diversion dams of quite different shape and size. Secondly, although the Kebar dam has a rather better finish to its faces, both dams otherwise exhibit the same type of construction; similar rubble masonry embedded in hard mortar with hydraulic properties. Thirdly, the fact that both dams used nearly identical devices for discharging the river during construction, and for drawing off water subsequently, again points to their having been built in the same era.

The most valuable evidence, however, comes from two structures built near the Kebar dam. One of the original cisterns which was supplied from the reservoir is of the same type of construction as the dam, and has moreover been identified as being of Mongol origin. Also about twelve miles from Kebar is a Mongol watch-tower, again featuring the same type of rubble masonry construction. There is little doubt then that the Kebar dam was built in the same period as the one at Saveh. Since the Kebar dam is better made and of more advanced design, it is likely that it is of slightly later date than Saveh (built at some time between 1281 and 1284) and not earlier. The reign of Ghazan Khan (who died in 1304) suggests itself as the most likely time in view of this ruler's efforts to develop irrigation. But whoever was responsible, the fact is that the Kebar dam, because it was a thin arch, is of great interest and importance and, together with the Saveh dam, places the achievements of the Mongols of Persia high on any list. This hardly bears out the decline in engineering with which they are usually associated.

The dynasty of Persian Ilkhans, that is to say the Mongol rulers, finally collapsed in 1349. For the next 150 years Persia was in a state of upheaval and for most of the time disunited. During this period Timur made his appearance and went some way to restoring unity, but only at the expense of continuous and savage fighting. His fourth son, Shah Rukh, who ruled from 1404 to 1447, also achieved a degree of unity and for a short period even succeeded in encouraging the arts and science. But in the latter half of the fifteenth century the country was once more in decline. During the period 1350–1500 the Kebar dam was silting up, and it is not known to what extent it continued to be used. It may in fact have been abandoned altogether perhaps owing to siltation, perhaps because of a general decline in agriculture during this troubled period. Otherwise there is nothing to report about Persian dams.

In 1501 the Safavid dynasty was founded and was destined to rule Persia for nearly 250 years. Religion was a powerful influence in the founding of the dynasty. Shah Ismail, who reigned until 1524, claimed descent from Ali, the fourth and last orthodox Caliph (656–61). This ancestry enabled Ismail to reunite Persia with Shi'ite Islam as the state religion. From 1587 to 1629, the Shah in power was Abbas the Great (Abbas I) and there is evidence of dam and other building under his influence. He turned his attention for instance to the maintenance and construction of roads, and this activity included the erection of a large number of caravanserais.

The one built near the Kebar dam, which is believed to date from the early seventeenth century, is almost certainly one of Abbas the Great's; in which case it is reasonable to attribute the heightening of the dam to him. Henri Goblot[26] has also drawn attention to a sixteenth-century dam at Quhrud near Kashan; it may be another arch dam, but beyond that nothing is known at present except the distinct possibility that it dates from the time of Abbas the Great.

Shah Abbas I's great-grandson, Abbas II, ruled from 1642 until 1667. Some dams can be attributed to his reign. There are, for instance, a number in the vicinity of Meshed, including the Band-i-Faridun, forty miles south-east of the city. It is of masonry construction, 120 feet high, 280 feet long and 24 feet thick at the crest. These impressive dimensions are more in keeping with the seventeenth century than, as has been claimed,[27] with the eleventh. Although heavily silted, the reservoir still supplies a small quantity of irrigation water. So also do two of the other Meshed dams at Turuq and Ghulistan.

Shah Abbas II's most important dams, however, are associated with the city of Isfahan. It stands on the banks of the River Zayandeh Rud, which was the source of its water supply. The Zayandeh Rud rises high up in the mountains to the west of Isfahan, and its source is very close to the source of another important Persian river, the Karun. About the middle of the sixteenth century Shah Tahmasp I conceived the ambitious idea of diverting Karun water into the Zayandeh Rud in order to increase the water supply to Isfahan.[28] But the construction of a connecting tunnel through a mountain ridge did not proceed very far before being abandoned owing, it is said, to the foul atmosphere in the workings.

At the end of the sixteenth century Shah Abbas the Great moved the capital of Persia to Isfahan and quite logically attempted to complete the project begun by his grandfather. The tunnel idea was abandoned in favour of an open cutting. It is reported that at times he employed up to 100,000 men on this mammoth undertaking, but to no avail. The high altitude produced impossible conditions of cold and snow in the winter. Yet the scheme remained an attractive one, and in the middle of the seventeenth century Shah Abbas II tried again. It is of some interest that Abbas II was advised in his attempt by a French engineer called Genest, and this is certainly one of the earliest cases of a European engineer being employed so far afield. It is possible that Genest attempted to cut the connecting channel by blasting the rock with gunpowder, but once more the mountains were the victors. Only 100 feet of the channel was constructed before the project was abandoned. Meanwhile, however, Genest had gone so far as to build a dam across the Karun in order to divert the river down the channel. The dam was 300 feet long and, for the time, very high, about 100 feet.[29] Such a height suggests that something more than mere diversion was intended. After all, as we have seen so often, a river can easily supply even a large canal from a dam no more than 10 to 20 feet high. It is reasonable to conclude, therefore, that Genest hoped to reduce the amount of excavation through the mountains by raising the level of the river as much as he dared.

The dam did not survive for long after the abandonment of the scheme, but the remains of his excavations did, and so too did the idea. In 1953 the Karkunan scheme, as it is now called, was completed.[30] The Kohrang tunnel, 9,250 feet long, connects the two rivers and at maximum flow delivers 1,050 cubic feet per second. The rubble masonry dam which was built to supply the tunnel is 33 feet high. The twentieth century statistics compare interestingly with those of the seventeenth century. By using a diversion dam three times as high, Genest would have reduced the length of the connecting channel by over 3,000 feet; and even if he had not succeeded in diverting as much water as the modern scheme does, he could presumably have achieved at least a few hundred extra cubic feet per second for the Zayandeh Rud. In short, modern engineering has completely vindicated an idea which in origin goes back 400 years.

The structure which probably represents the peak of Safavid civil engineering is in Isfahan. The Pul-i-Khadju (Pl. 21) was built by Shah Abbas II and represents the ultimate achievement in the long Persian tradition of bridge-dams.[31] Apparently there was a dam across the Zayandeh Rud at the same point at some earlier date. Nothing is known of this structure except that Abbas II's dam was built on top of its foundations. The dam raises the river level 20 feet and is 462 feet long. The total thickness of over 100 feet is deceptive, because it represents the width of dam necessary to support the bridge rather than the width needed to resist the water-pressure. Both dam and bridge are made from dressed stone blocks accurately fitted together to give a structure of great strength and durability. The dam is pierced by sixteen narrow openings extending to a depth of some ten feet below the crest. Normally the river discharges through these openings each of which is fitted with sluice gates. In times of excessive flow when the sluices cannot cope, the surplus water passes over the crest of the dam and through the lower tier of the bridge arches. On the air face, the masonry is stepped from the crest of the dam down to tail-water level, while from above the dam water was, and still is, taken off for irrigation around the city.

This is not the place to discuss in detail the magnificent bridge which stands on Abbas II's dam. Suffice it to say that it has two storeys, the upper one comprising two sets of pointed-arch enclosures flanking the roadway which is carried on the lower set of arches. Rising above the roof of the upper storey at the centre of the bridge is an octagonal pavilion, and there are smaller pavilions at each end. Both as a bridge and as a meeting-place for the populace, the Pul-i-Khadju ranks with Old London Bridge and the Ponte Vecchio. Architecturally it is probably the peer of both, and indeed it is thought by many to be one of the most beautiful bridges anywhere in the world. But whatever the merits of the superstructure, the dam beneath is the important thing to note here. Although unspectacular, it is well made and entirely adequate; and it ends our study of Persian dams.

4

The Moslems

THE MOSLEM ERA is conventionally taken as beginning in A.D. 622, the year in which the Prophet was forced to flee from Mecca to Medina. Muhammad died in 632 and his teachings of the previous twenty years triggered off a conquest which in terms of speed and size is impressive by any standards. Before the end of the first decade of the eighth century Moslem rule extended from the Indus valley to the Atlantic coast of North Africa, and shortly afterwards the Moslems added Spain, and for a short time even part of what is now France, to their possessions. This contact between Europe and the Islamic empire was to be of great consequence to both parties, not least in technology.

It must be stated, however, that the history of Moslem technology has not been given anything like the attention that it deserves or that is necessary if its origins, its development and its effects, particularly on European techniques, are to be fully understood.[1] Historians of civil engineering have almost totally ignored the Moslem period, and in particular historians of dam-building, such as there have been, either make no reference to Moslem work at all or, even worse, claim that during Umayyad and Abbasid times dam-building, irrigation and other engineering activities suffered a sharp decline and eventual extinction. Such a view is both unjust and untrue.

The sources of Moslem expertise in the related arts of dam-building and irrigation are patently obvious. One need do no more than point out that nearly all the dams, irrigation works and water-supply works which have been discussed in preceding chapters were built in an area which, by the middle of the eighth century, was under Moslem control. Undoubtedly the example of Roman engineering was especially prominent because of the scale on which they worked and because their influence was relatively recent. The testimony of the Moslem geographers, later than the eighth century it is true, confirms the predominance of the Roman exemplar. What proportion of these hydraulic works, whether of Roman or some other origin, were still in use and what proportion had been allowed to decay through misuse or lack of maintenance is impossible to say. It is merely important to note here that those which had survived could, whatever their state of repair, serve as an example. Spain, North Africa, Egypt, Syria, Arabia, Persia, Mesopotamia and Armenia constituted a large reservoir of hydro-technology which the Moslems inherited.

In the process of assimilation and development which occurred, the attitude of the Arab conquerors to their new possessions is significant. The structure of civil and religious administration was to a large degree left undisturbed, and the traders, artisans and agriculturists among the conquered peoples, although of inferior status compared with their Moslem masters, were allowed to pursue their occupations as before. Preservation of established technological skills was thus ensured.

In Arabia there were already dams in the pre-Muhammadan period—the Marib dam, perhaps those at Adraa and Adschmaa, dams in the Najran and those around Taif built by Jewish irrigators. And it is near Taif that we find the first Moslem dams. Several masonry irrigation dams have survived of which one is still in a good state of repair.[2] It is to be found twenty miles to the east of Taif and was built to conserve soil and water in the Romano-Nabataean fashion. The dam is crudely constructed of large boulders and is 25–30 feet high and about 250 feet long. Today the structure has little ability to hold water, and its rough construction seems to indicate that it never did. Of more interest than the dam itself, however, is the inscription which the builder cut into the rock at one end of the dam. It is believed to be the oldest Kufic inscription known and reads, in translation:

> This dam belongs to Abdullah Muawiyah, Commander of the Believers. Abdullah bin Sakhr built it, with the permission of Allah, in the year 58 [A.D. 677/78]. Allah, pardon Abdullah Muawiyah, Commander of the Believers, and strengthen him, and make him victorious, and grant the Commander of the Believers the enjoyment of it. Amru bin Janab wrote it.

The dams near Taif are the only ones known to have been built by the Moslems in Arabia. For more of their dams one must look elsewhere.

Damascus was established as the capital city of the Umayyad caliphs in 661. For a hundred years it was the centre of the Islamic world and became a splendid city. Its water-supply system, dating back to Roman times, was improved and enlarged by the Moslems who paid particular attention to the construction of regulating dams on the River Barada. Parts of this system have survived to the present day.

After Egypt the most important agricultural area in the Islamic empire was Iraq, especially following the collapse of the Umayyad caliphate in 750. In 762 al-Mansur founded the capital city of the Abbasid caliphs at Baghdad, in the centre of a region which was already extensively irrigated by the Euphrates, Tigris and other rivers. It was probably during Sassanian times that the Euphrates gradually retired from its eastern course past Hilla and Diwaniyah and, starting at Musaiyib, had cut a new channel past Hindiya and Kufa, somewhat along the line of Alexander the Great's Pallacopas canal. So when al-Mansur came to Baghdad the Euphrates was following much the same course that it does today. During Abbasid times this vagrant river was on the move once again, so that by the end of the twelfth century it was once more discharging almost all of its flow down the original eastern branch.

The Tigris between Nimrod's dam (see Chapter 1) and Baghdad was, in the eighth century, still flowing in its western bed, as the remains of numerous and once prosperous Moslem towns prove. Below Baghdad the Tigris took a more westerly course than it does today. From Kut it flowed past Wasit and then into the Great Swamp, a vast marshy region which at its northern corner also received the Euphrates. The Kut–Wasit course was used throughout Abbasid times right down to the sixteenth century. Thus the ancient easterly loop of the Tigris, deserted in Sassanian times, was a dry valley. And yet the Moslems dammed it. The geographer Ibn Rustah mentions a dam at Madhar designed to create a navigable backwater fed from the Tigris' main stream at Kurnah. Details of this early dam (about A.D. 700) have not survived.

From the outset Baghdad's founder, al-Mansur, applied himself to the enlargement and extension of the ancient irrigation works in the region. South of Baghdad were five major irrigation canals: the Nahr Isa, the Nahr Sarsar, the Nahr Malik, the Nahr Kutha and the Shatt-an-Nil (Fig. 9). All these canals appear to have been reworkings of older installations, and by enlarging them the Moslems not only obtained more irrigation water but also established navigations between the 'two rivers'. The re-excavation and realignment of the Nahr Isa resulted in an especially large canal which irrigated a large area of land, provided passage for large ships and was spanned by ten big masonry arch bridges.

Because its mouth was so close to Baghdad, the Nahr Isa was a particularly important canal about which a good deal is known from the pen of Ibn Serapion, a Moslem geographer who wrote early in the tenth century.[3] In typical fashion the Nahr Isa fed a series of parallel secondary channels which in turn supplied numerous small transverse canals. This was perhaps the most important irrigation technique which the Moslems inherited and then developed for themselves. Of considerable importance was the fact that each plot of land, in addition to being well supplied with water, could also be properly drained by running away excess water to a lower level. Thus the crops were not only effectively watered but the water was efficiently used.

The Nahr Isa's irrigation was controlled by small dams of unknown form. Of greater interest is the system of dams used to regulate the Nahr

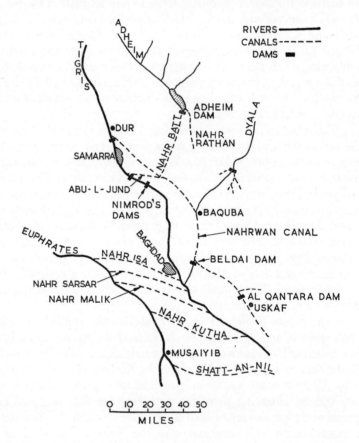

Figure 9   The system of dams, canals and rivers in the vicinity of Baghdad during the period of the Abbasid caliphs.

Isa's intake on the Euphrates, which, according to al-Khatib (1002–71), were at a place called Kubbin. The size of the canal was undoubtedly of great benefit to irrigation and shipping, but only at the risk of increased flood damage in the event of the Kubbin dams being overtopped or destroyed. This price was paid more than once. The dams failed in 942 causing severe flooding in the south-western suburbs of Baghdad, and many citizens were evacuated. The same thing happened again during the reign of Musta'sim (1241–56), the last Abbasid caliph, but in neither case is the cause of failure known in detail.[4]

In Abbasid times there seem to have been three dams across the Tigris north of Baghdad (Fig. 9). One was the ancient Nimrod's dam; the second was the diversion dam near Dur which fed the Nahrwan canal; and the third dam, just to the south of Samarra, was built by Harun-al-Rashid to supply the Abu-l-Jund, a canal which drained into the main stream of the Nahrwan canal. Abu-l-Jund has the picturesque meaning 'Father of the Army', a reference to this canal's use in irrigating the crops for Harun-al-Rashid's supplies of food to his army.[5]

The finest Moslem dam in the vicinity of Baghdad was built on the Adheim river (Fig. 9). Perhaps as late as the Sassanian period the Adheim was a tributary, not of the Tigris but of the Dyala, its confluence with the latter being somewhere near Baquba. The Adheim itself is a flashy river of small average flow, and in order to increase its potential as a source of irrigation, the Moslems took the trouble to dig feeder canals from the more northerly Lesser Zab river. This achieved, it was then necessary to build a diversion dam on the Adheim at the point where it leaves the hills called Jebel Hamrin.[6] The dam's remains are shown in Figure 10.

The main body of the dam is a masonry wall 575 feet long which at the western end turns through a right angle and continues for 180 feet to form one bank of the canal called Nahr Batt. The dam has a maximum height of something over 50 feet, but this rapidly reduces towards the sides. In fact for the first 150 feet at the eastern end, the dam is only 13 feet high. The cross-section of its central portion has a neat trapezoidal profile, 10 feet thick at the crest and 50 feet thick at the base. The water face is vertical and the air face is built to a uniform slope with the masonry stepped. The drawing clearly suggests that the dam was built of cut masonry blocks throughout, and these, it seems, are connected with lead dowels poured into grooves. This was a not uncommon Moslem technique and in the Adheim dam was apparently used as a complete alternative to mortared joints. The alignment of the structure is not straight, and this reflects an attempt, as usual, to utilise the natural shape of the site as advantageously as possible.

The Adheim dam, a solid piece of construction of impressive size, was built to feed the two irrigation canals shown in Figure 9. The Nahr Batt drained eventually into the Nahrwan canal very close to the Abu-l-Jund; the mouth of the Nahr Rathan may have been somewhere on the Dyala, but this is not certain. Because the Adheim's flow had been boosted by water diverted from the Lesser Zab it is thought possible that the dam served to provide a degree of flood-control. To what extent this purpose was served by a low-level sluice is impossible to verify now that the central portion of the dam is in ruins. It seems fairly clear, though, that any

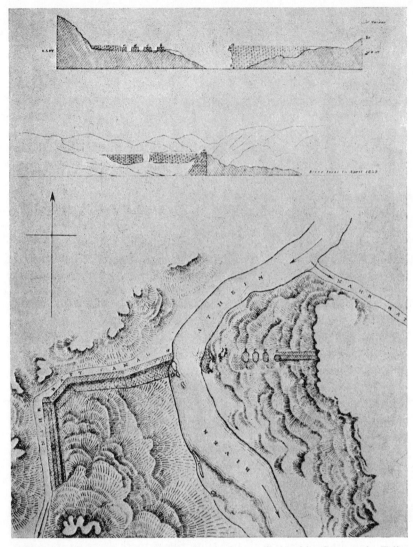

Figure 10 The ruins of the Adheim dam in Iraq as depicted by Commander Felix Jones, R.N., in the middle of the nineteenth century. (Courtesy of the Institution of Civil Engineers.)

overflow was discharged straight over the crest. The date of the Adheim dam's failure is not known, but since this event the river has flowed unimpeded into the Tigris.

Perhaps the demise of the Adheim dam and its attendant irrigation works were part and parcel of the general decline and eventual collapse of Iraq's irrigation which occurred in the later centuries of Abbasid rule.

The failure of the Sassanian Beldai dam is probably another example; while the Abbasids' reconstruction in the eighth century of the large regulating dam of Al-Qantara was rendered useless in the twelfth century by massive siltation quite beyond the administration's ability to control. Even more disastrous was the later caliphs' inability to check the reversion of the Tigris to its ancient easterly course above Baghdad. This shift, which wrecked the upper reaches of the Nahrwan canal, must have been well advanced during the reign of Mustansir (1226-42) because we know that he tried to build canals to water the land deserted by the river. Implicit in this situation is the probability that Nimrod's dam had ceased to be effective or was already a ruin; and there was no one to engineer a solution.

At the time when the Abbasid rulers of Baghdad were powerful and committed to the construction and maintenance of dams and canals, the Euphrates–Tigris–Nahrwan irrigation works reached their maximum development. A decline of authority, particularly in the eleventh and twelfth centuries, allowed the dams to deteriorate and collapse, the canals to be choked beyond recall, and salination of the land to complete the desolation from which Iraq is only now at last beginning to recover.

Outside Iraq some of the most elaborate irrigation systems which the Moslems inherited from the Sassanians were the ones at Shushtar, Ahwaz, Band-i-Kir and Dizful. All are referred to by several different Moslem geographers who make it clear that the dams and canals were in continuous use in Abbasid times. Repairs and reconstructions were numerous, however, and this is especially evident from the pointed arch bridges which can still be seen at Shushtar and Dizful, partial replacements of the original Romano-Sassanian work.

To the Shushtar system the Moslems added a dam called the Pul-i-Bulaiti.[7] There is abundant evidence that the Moslems used the waterwheel extensively for irrigation and milling, and the Pul-i-Bulaiti on the Ab-i-Gargar was built for the latter purpose. The mills were installed in tunnels cut through the rock at each side of the channel, the dam, of course, being built to provide the necessary head of water. It is one of the earliest of hydro-power dams but by no means the only Moslem venture of this type. Another example is the bridge-dam at Dizful which was used to power 'a great water-wheel working a mechanism which raised water 50 ells [190 feet no less] and thus supplied all the houses of the town'.[8] Mills were also driven from the dam at Ahwaz, and for this we have the word of Mukaddasi (tenth century) who says, 'The canals [from above the dam] separate at the highest part of the town and rejoin lower down at a place called Karschnan. From there ships can pass to Basra. And they have wonderful mills on the water.'[9]

In the nineteenth century the channels in which the water-wheels were

D

located could still be seen. Some of them, or rather later versions of them, were still in use for corn-grinding but not for crushing sugar cane, one of their principal uses in Moslem times.

In general the dams and canals on the rivers Karun and Ab-i-Diz suffered a similar fate to those around Baghdad. The decline in authority of the Abbasid rulers allowed the systems at Band-i-Kir, Ahwaz and Dizful to go out of use, and the towns themselves fell into decay. Only the Shushtar dams have survived in anything like a useful condition.

From 945 until 1055 the real power in Abbasid Iraq and Persia lay with a dynasty of rulers called the Buwayhids. Their most famous ruler was Adud-ad-Dawlah (949–83), a man whose power and influence are well reflected by his engineering works: building mosques, palaces and hospitals; constructing immense fortified walls round Isfahan; deepening and widening the ship canal between the Tigris and Karun, and rebuilding the bridge across the river at Ahwaz. And he was also a dam-builder.

One of the most important Moslem towns in the southern province of Fars was Istakhr, the Persepolis of ancient times. Part of the defences of Istakhr consisted of three large fortresses located in the hills to the north-west. In the event of a siege these forts were supplied with water from a great reservoir covered by a roof carried on columns. Ibn-al-Balkhi (early twelfth century) has left a valuable account of this work.[10]

> [For the castle of Istakhr] Adud-ad-Dawlah built a mighty tank, which is known as the Hawz-i-Adudi. It was constructed in a deep gully, down which the stream that passed by the castle flowed. First, Adud-ad-Dawlah with boardings closed the end of this gully, making the like of a great dam, and next inside this he set cement [in forms] with wax and grease laid upon kirbas-stuffs, with bitumen, bringing the whole structure to the upper level all round, and afterwards when it had settled down firmly nothing could be stronger. Thus was the tank made, and its area was a qafiz [a square of 144 ells] all but a fraction, being 17 feet in depth, wherefore if a thousand men for a whole year were to drink therefrom, the water level would not sink more than a foot. Then in the middle part of the tank they built up twenty columns of stone, set in cement, on which they rested the roof that covered the tank. Further Adud-ad-Dawlah, besides this tank, built other water tanks and cisterns.

The size of the dam is not stated, but one gets the impression that a suitably narrow part of the 'deep gully' was chosen so that quite a small dam gave the required reservoir capacity. The dam, it would seem, was made of wood. It is difficult to know what to make of the way in which the other materials were used. The likely explanation is that they constituted a waterproof lining on the water face of the wooden dam.

Certainly some sort of facing would be needed with a wooden dam unless the carpentry was of an exceedingly high standard. 'Kirbas-stuffs' was a heavy cotton fabric with a texture like that of thin canvas. It would seem then that the dam was faced with cloth impregnated with grease and wax, and the whole sheet held in place by a layer of bitumen. Why the bitumen would not have been satisfactory by itself is a mystery.

If the above interpretation is correct, then the dam of the Hawz-i-Adudi represents one of the most extraordinary ever built so far as materials are concerned. What is more Ibn-al-Balkhi presents us with a rare reference to the use, in Abbasid times, of bitumen as a building material, even if it was only as a waterproofing agent. The reservoir was covered in order, presumably, to prevent evaporation losses, an important requirement in a siege, and also to keep the water clean. Ibn-al-Balkhi says that one thousand men would use one foot depth of water in a year, a generous allowance of about five gallons per man per day.

Adud-ad-Dawlah's most impressive achievement as a dam-builder was the Band-i-Amir. It was built in 960 on the River Kur in the province of Fars some sixty miles upstream from Lake Bakhtigan. The Band-i-Amir was described by both Mukaddasi in 985 and Ibn-al-Balkhi in 1107. Here is the latter's account.[11]

> Next comes the Adudi Dam, the like of which, as is well known, exists nowhere else in the whole world. To describe it, it must be known that the Kirbal District [which lies round and about] originally was a desert plain without water. But Adud-ad-Dawlah seeing this opined that if a dam were built here the waters of the River Kur would work wonders on this desert land. He therefore brought together engineers and workmen, and expended great sums of money to make side canals to lead off the waters of the river from the right and left bank. Then he [paved the river-bed], above and below the dam, with a mighty weir [shadurwan] constructed of blocks of stone set in cement. Next he built the dam itself with [stones set in] tempered cement and sifted sand, so that even an iron tool could not scratch it and never would it be burst asunder. The summit of the dam was so broad that two horsemen could ride abreast across it without the water touching them, for to carry this off sluices were made. Thus, finally, the whole of the district of Upper Kirbal received its irrigation by means of this dam.

Mukaddasi, who wrote only twenty-five years after the dam was built, agrees closely with the above account and adds two valuable pieces of information: firstly, that the dam was built of 'stones set in mortar, reinforced by iron anchors set in lead', and secondly, that 'ten water mills were built close to the dam'.[12]

Plate 22 is a downstream view of the dam as it is today. It is interesting that neither Mukaddasi nor Ibn-al-Balkhi mentions the pointed arch bridge and in fact both state that two horsemen could ride across the *crest* of the dam. Evidently, then, the bridge is a later addition, and in its final form the Band-i-Amir closely resembles the Romano-Sassanian structures at Shushtar, Dizful and Pa-i-pol.

The dam itself is some 30 feet high and 250 feet long. Details of the water face are not available because of heavy siltation, but the air face can be seen to be built with a slope of about 45 degrees. This face is divided into smooth sections, which coincide with the bridge arches, and stepped sections, which coincide with the piers. In times of flood, overflow is discharged over the dam's crest and through the bridge arches.

Apparently the Band-i-Amir is built of masonry blocks throughout and does not make use of a rubble masonry core. Once more we find iron bars set in lead being used to connect the blocks as in the Adheim dam and those on the Tigris. Both writers agree that the stones were set in mortar, and this, in addition to binding the whole construction together, would also have served to make the dam watertight. The use of 'tempered cement and sifted sand' illustrates that the engineers were aware of the need for careful preparation of their mortar. 'Even an iron tool could not scratch it,' writes Ibn-al-Balkhi, in which case an excellent mortar must have been obtained. It is evident that the Band-i-Amir was a very thorough and solid piece of work, and it is not at all surprising that it has had such a long and useful life.

At the left-hand end of the dam can be seen the remains of masonry walls built at right angles to the line of the structure. They appear to be the outlet sluices mentioned by Ibn-al-Balkhi and were used to maintain a supply of water to other irrigation dams further downstream. Perhaps the outlet channels also contained the ten water mills to which Mukaddasi refers. It is interesting that in addition to grinding corn the wheels of these mills were also used to raise water for irrigation. It is becoming increasingly apparent, then, that the use of dams to provide hydraulic power, albeit in a very elementary form, was a regular feature of Islamic hydraulic engineering.

The Band-i-Amir is only a diversion dam feeding numerous canals on both banks; it has no storage capacity to speak of. Nevertheless in its hey-day it irrigated a large tract of the Kur valley, Mukaddasi mentioning 300 villages. Today, however, the dam is increasingly unable to support irriga-tion because of silt. This old enemy, the bane of all irrigation systems, has not only choked many of the canals but, near the dam, the river as well.

Nearly fifty miles upstream from the Band-i-Amir stands the ancient

Band-i-Naseri, a dam of Achaemenian origin. Ibn-al-Balkhi describes it as follows:[13]

> In this part of the stream they had in former days erected a dam in order to secure a sufficiency of water to irrigate the lands, but in the times of disorder when the Arabs overran Persia this dam fell to ruin, and all the districts of Ramjira went out of cultivation. In recent years the Atabeg Chauli has rebuilt this dam, and the country round has again been brought under cultivation.

The Atabeg Chauli was the Seljuk governor of the region from 1098 until 1116 and was responsible for a good deal of dam-building and re-construction. His work on the Band-i-Naseri was not effective for very long, however, because about A.D. 1200 it had to be rebuilt again. There have been several reconstructions since that date but it still continues in use. Its record of 2,500 years of service to irrigation, even allowing for long periods of disuse, is an impressive one.

Downstream from the Band-i-Amir, and situated therefore between the dam and Lake Bakhtigan, are five more ancient irrigation dams, none very large.[14] The first of them is the Achaemenian Band-i-Feizabad, which is mentioned by Ibn-al-Balkhi under another name: 'The Band-i-Qassar had been built of old to water the district of Lower Kirbal, and it too had fallen out of use; but the Atabeg Chauli has likewise restored this to working order and some distance below it the River Kur flows out into the Lake of Bakhtigan.'[15]

The dam's name is particularly interesting; it means the 'Fullers' Dam'. Although it was built for irrigation and is still used for this purpose, the name suggests that it was also used for fulling, and this is not unlikely. Fulling has been practised for centuries in Persia as other names bear witness: at Samarkand, for instance, a canal called the 'Fullers' River' was used to drive water-wheels for fulling mills.

Ibn-al-Balkhi's reference to the Band-i-Qassar is important because he regards it as the last one above Lake Bakhtigan. Thus the other four dams on the Kur must post-date his book of 1107. It is also of interest that although Ibn-al-Balkhi mentions repairs to the Band-i-Naseri and the Band-i-Qassar by the Atabeg Chauli, he does not say that this ruler also renovated the Band-i-Amir and perhaps built the bridge; for this information we are dependent on other later writers. Thus it would appear that the Atabeg restored the first two dams before 1107 and the Band-i-Amir after 1107 but before 1116. In this instance the chronological data are more precise than is usual.

It is clear from the above discussion that, although the Kur is not a large river, it has for 2,000 years at least been extensively utilised for irrigation, and that the system became most fully developed in the Abbasid period.

For other Moslem dams in the eastern part of the Islamic empire, our information is disappointingly vague, and a brief outline will suffice.[16]

Zaranj, the capital of the province of Seistan, had been founded in Sassanian times following the failure of a dam on the River Helmund which forced the people to move. Various Moslem geographers indicate that under Abbasid rule Zaranj was irrigated by six dams on the Helmund near the point where this river enters Lake Zarah. Some or perhaps all of these structures may have been Sassanian in origin. The Moslems added numerous water-wheels to the system, but then in 1383 the city and its irrigation works were destroyed by Timur. A similar fate befell the Band-i-Rustam, another Moslem dam on the Helmund which irrigated a large area south of the city of Bust, an important place which was near the modern Girishk.

In the province of Khurasan the Murghab river was dammed by the Moslems to irrigate the city of Great Marv. Even more significant was the dam at al-Jurjaniyah, south of the Aral Sea. For centuries the River Oxus flowed from the mountains of Pamir to a point about 100 miles south of the Aral Sea and here it turned west and crossed Turkmenistan to the Caspian Sea. Some time before the tenth century the Oxus, a large river, abandoned its Caspian Sea outlet and took the much shorter route to the Aral Sea. Thus for most of the Abbasid period the Oxus followed more or less its modern route. Along the last 300 miles or so of its course the Moslems built numerous diversion dams to feed canals, and one of these canals was nothing less than the deserted river-bed to the Caspian Sea. At al-Jurjaniyah the old river-bed was dammed to control the water level and terminate the canal.

In 1220 the armies of Jenghis Khan smashed the dam and swamped and ruined al-Jurjaniyah. But there was also a more dramatic long-term effect. The destruction of the dam allowed the Oxus once more to flow down its ancient course to the Caspian Sea. So complete was the reversion that by the fifteenth century the Aral Sea, deprived of water from the River Oxus, almost dried up. The failure or destruction of a dam can have considerable and unpredictable consequences.

One of the River Euphrates' tributaries in its upper reaches is the River Hirmas, and there are references in the Moslem geographers to a Byzantine dam, 'a masonry wall clamped and with leaden joints', on the Hirmas. Further downstream, about 100 miles north of Nusaybin, the Moslems, in the tenth century, built the Sukayr al-Abbas, a large dam which split the Hirmas into two streams. The eastern branch, which was the larger, was called the River Tharthar, and after flowing for more than 200 miles in a south-easterly direction it joined the Tigris just above Samarra. A certain interest attaches to the Tharthar because it was, for a few hundred years, the only tributary of any size on the western bank of the Tigris.

The River Tharthar no longer exists; in fact when the geographer Yakut wrote in the thirteenth century it was already virtually dry except when the Hirmas was in flood. This surely indicates that the Sukayr al-Abbas dam had failed or was at least out of order like so many other hydraulic structures of this area in the thirteenth century.

The dam-building achievements described in this chapter so far span a period of about 600 years, from the middle of the seventh century to the middle of the thirteenth. It is true that a great deal less detail is available than one would like; many dams are known only by name, and their locations are anything but precise. However, their inclusion in the story has been worth while, if only to do justice to Islamic engineers and at the same time to correct the historical record.

It is true, of course, that the centre of activity shifts from place to place, reflecting to a large extent the way in which the real power in the empire was held by different rulers at different times. Also there was an eventual decline over the last two centuries of Abbasid rule, a decline which in many places the Mongol invaders did little to check or reverse, Persia, as we saw in the last chapter, being the notable exception. Nevertheless, the Abbasid engineers in their prime were able to make their mark on the history of engineering. Limitations of space prevent a full discussion of their other civil engineering works, particularly the construction of numerous cisterns and tanks, frequently very large, for water supply; aqueducts and aqueduct bridges; qanats; and occasionally 'inverted siphons' in pipe lines. Nor can we spend any time on Moslem water law which, when examined thoroughly, may well demonstrate a Moslem contribution to hydraulics.

In general Moslem dam-building lacks originality so far as the structures themselves are concerned. To a large extent this was a result of their rebuilding and restoring older works rather than constructing new ones, an inevitable course of action when it is remembered that the Moslems, as successors to the Romans, Byzantines and Sassanians, inherited so much that was suitable for reconstruction. But even in dams of wholly Moslem origin, like the one on the Adheim or the Band-i-Amir, everything was done according to the established formula. Masonry was always the material of construction for the obvious and good reasons that had predominated in earlier periods: it is both heavy and durable.

With rare exceptions the stability of ancient dams depended on their having sufficient weight, something which early dam engineers knew very well they could achieve by using masonry in a profile of ample, and by our standards excessive, dimensions. In the absence of any form of quantitative structural analysis such a practice was inevitable, and in periods when materials and labour were cheap it was perfectly sound engineering to achieve stability in this way. The development in modern times of methods of designing slender dams which are structurally more efficient merely reflects the fact that materials and labour are no longer cheap commodities, and also that the time available for the construction of a dam is often limited.

The other advantageous property of masonry, namely its ability to resist the erosive effects of running water, was particularly valuable to Moslem engineers. Almost without exception their dams were intended to raise and divert some proportion of a river's flow into canals and discharge the rest over the dam's crest, this action being quite automatic. But in dams such as the Band-i-Amir where sluices were used to control the discharge, masonry construction was even more essential to cope with higher water velocities.

The Moslems' adherence to traditional methods of construction even extended to the Romano-Sassanian and Byzantine practice of using mortar, probably hydraulic, for strength and watertightness, and sometimes lead and iron connections both for strength and as an aid to location of the blocks.

Even though the Moslems contributed nothing new to the form of dams, in one respect certainly, and possibly in another, they were innovators. Already we have seen that they pioneered the use of dams to drive water-wheels, these devices being widely used in Islam to raise water, to grind corn and to power fulling mills. It is interesting that while the Moslems were pursuing this development in their part of the world, Christians were engaged in exactly the same pursuit in Northern Europe.

The second possible Moslem innovation concerns surveying. In order to locate suitable dam sites and to lay out complicated canal systems, the civil engineer is required to carry out surveys. It is possible but as yet unproven that Moslem engineers made improvements to this art.[17] Surprisingly, perhaps, they did not take over the surveying instruments of the Romans but instead, for much of their levelling work, used simple plumb bob devices. But they also levelled with the astrolabe, an instrument which they brought into general use and introduced into Europe. In conjunction with their basic and important developments in trigonometry, it would have been possible also for Moslem engineers to triangulate with the astrolabe and in general to practise surveying on a larger and more accurate scale than their predecessors. Pending more research, however, this can be nothing more than a hypothesis. What is certain is that Moslem engineers laid out some vast networks of interconnecting canals, often between rivers, located dams with precision and were clearly competent surveyors.

With these exceptions, there is nowhere in Moslem dam-building any radical departure from or improvement on Roman, Byzantine or Sassanian practice. But even though they were not innovators in dam-building they were the next most important thing, preservers. They were able to transmit something of the techniques of earlier times to their contemporaries and successors. Nowhere is this transmission of technical ability more obvious than in Spain.

Roman rule in Spain was replaced during the fifth century by that of the Visigoths. Their domination of Spain lasted until the eighth century, but during this period there is no evidence of dam-building and very little evidence of any other engineering works. However, it is possible that some Roman dams, water-works and irrigation systems were kept in more or less working order. That irrigation was practised by the Visigoths is clear enough because their book of laws, the *Liber Judiciorum*, prescribes penalties for the theft of water.

In A.D. 710 a small Moslem force reconnoitred southern Spain, a large-scale expedition followed a year later, and by A.D. 712 the Moslem conquest was virtually complete but for the northern provinces of Galicia, Asturias and Cantabria. These remained in Christian hands after the half-hearted Moslem raids of 718 had been repulsed. Had the Moslems persevered and conquered the north as well, the course of Spanish history might have been very different. As it was, the kingdom of Asturias provided the basis of the Reconquest, although the territory captured by the Moslems in seven years was to take the Christians seven centuries to recover.

Although it is a fascinating question and one requiring a great deal of study, it is not proposed to discuss in general the role of the Moslems as transmitters of technology from the East to Western Europe. Spain was, however, the most important link between the two, and particular cases of technical contact, such as paper-making, steel manufacture and sericulture are well known. Books on technology undoubtedly found their way to Spain, too, although their effects are at present unknown. The Moslems also introduced many new crops, of which four are especially important: rice, oranges, cotton and sugar. To grow these irrigation was essential, and it is the development of this aspect of technology which concerns us here.

There is plenty of evidence, some of which will be mentioned, that the Moslems brought irrigation techniques to Spain from the Middle East and thereby laid the foundations of the agricultural prosperity which is one of the most impressive characteristics of Moslem Spain. Nothing so elaborate and efficient had been seen before in Europe. Details of how the transmission of irrigation techniques proceeded are not clear, but a few indications can be given. Before A.D. 750, while still an integral part of the Islamic empire, Spain was systematically colonised by large numbers of soldiers from Syria, Egypt, Persia and Iraq, men who had made up the bulk of the army of conquest. Because these people came from the valleys of the Orontes, Nile, Tigris and Euphrates, the area around Damascus and from the Negev, it is entirely reasonable to suppose that they brought with them the centuries-old irrigation techniques with which they were so familiar. The fact that they settled in the great river valleys of southern Spain gave them ample opportunity to introduce these techniques to that country.

Large numbers of civilians drifted west, too, and this migration probably continued after 750, that is after the split between Umayyad Spain and the Abbasid empire in Asia and North Africa. In spite of political and geographical separation, technical and cultural contacts were maintained. There was a steady interchange of scholars, teachers and students, while Western Moslem geographers and historians travelled widely throughout the Islamic world and beyond. Commercial contacts as well were probably effective in diffusing technology across the length and breadth of the Moslem empire. No less impressive than the astonishing variety of goods exported and imported by Spain over a long period are the distances over which some of them travelled.[18] China and India, for instance, figure prominently in the picture.

The Moslem domination of Spain was centred on the south, in al-Andalus. To Arabians, Syrians, Persians and Berbers it was much like their homelands: the environment was familiar and manageable and they knew how to make the country fertile. Every river of any importance from the Ebro right round to the Guadalquivir was dammed for irrigation, and sometimes for water supply and water power as well.

Cordova was the capital city of Moslem Spain for nearly 500 years, and it is here, on the Guadalquivir, that we find what is probably the oldest surviving Moslem dam in Spain; it stands just below the Puente Romano. As Plate 23 shows, the line of the dam follows a zig-zag course across the river so that its total length is about 1,400 feet even though the river is only of the order of 1,000 feet wide. This shape undoubtedly indicates that the builders were aiming at a long crest in order to increase the dam's overflow capacity.

Today the weir is very dilapidated and it is not easy to decide exactly how it was built; but rubble masonry is everywhere in evidence, and at one time this may have carried a cut masonry facing. In its prime the dam was probably about 7 or 8 feet high and 8 feet thick. At each of the three downstream points of the zig-zag stand the remains of mill houses. These are mentioned[19] by al-Idrisi (twelfth century) who says that each one contained four water-wheels; details are not given, however.

The dam served a second function apart from milling. At the Cordova end it is built into a natural bank close to the long and massive wall which protects the city from river floods. Here the Moslems installed a huge noria which was driven by the water raised above the dam. The noria lifted water from the Guadalquivir to the level of an aqueduct, still visible, and this then carried it into the city. Thus the old Cordova dam met the twin needs of milling and water-supply; and as a bonus it has for 1,000 years protected the Puente Romano's piers from scour.

The golden age of Umayyad Spain lasted ninety years, from 912 to 1002. The dams which are about to be discussed all date from this period and can be conveniently referred to as 'tenth century'. The origins of some of them may go back to the ninth century or earlier, but it was during the peaceful and energetic reigns of the tenth-century caliphs that public works prospered most markedly. Then it was that the irrigation systems were at their maximum extent and greatest efficiency. The tenth-century rulers encouraged these developments, provided capital from the state's revenue for construction and maintenance, and introduced the administrative machinery which was required to ensure efficient and equitable operation. Once more we find that the construction and operation of engineering works on a grand scale is a feature of powerful and stable government, utilising its authority to raise money and direct manpower in a time of peace.

The 150-mile-long River Turia, which flows into the Mediterranean at Valencia, was known to the Moslems as the Guadalaviar, or White River. In the tenth century there were many small dams or *azuds* on the river, of which eight, all in Valencia, are of particular interest.[20] (In passing it is very important to note the Spanish word *azud*. It is but one of very many modern irrigation terms taken directly from Arabic and certain proof of the Moslem influence on Spanish technology.)

These eight dams are spread over six miles of the river's course. Each feeds a single main irrigation canal, of which there are four on each side of the river. The irrigated area adjacent to the river covers nine square miles, but some of the canals carry water much further, particularly to the Valencian rice-fields. These, of course, were established by the Moslems, and they continue to be one of the most important rice-producing centres in Europe. Because all the Valencian dams are so similar in size, shape and design, only one will be discussed.

The dam of Mestella (Pl. 24, Fig. 11), the fifth in the series, is 240 feet long and 7 feet high. The water face is vertical, the air face stepped, the crest is 4½ feet wide and the base thickness 18 feet. The core of the dam consists of rubble masonry and mortar, and the structure is faced with large masonry blocks with mortared joints. On some of the other dams iron connectors are used as well as mortar. At its left-hand end the Mestella dam rests against a heavy masonry abutment, as do several of the other *azuds*. This wall, which is three feet higher than the crest of the dam, was designed to guide overflowing water away from the river bank to prevent scour of the foundations. But nowadays this presents no problem. More than half the structure is silted up right to crest level, so that, although the full length of the crest was originally designed to discharge the river, at present it is forced to flow in a channel of much-reduced width and spills over the weir only at its eastern end.

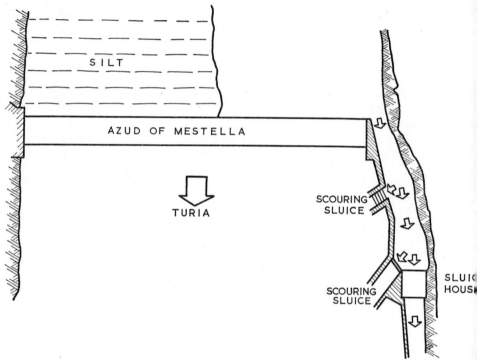

Figure 11   Plan view of the *azud* of Mestella in Valencia.

At the same end the dam abuts on to a masonry wall which extends downstream some seventy feet and is everywhere the same height as the dam and similarly built. Between this wall and the river bank a proportion of the Turia's flow is directed to the mouth of the irrigation canal. Two sluices are built into the wall, one half-way along the other near the canal mouth. They served two purposes: during normal operation they were used as escapes to allow surplus water to drain back into the river; and occasionally they would both be opened to their full extent in order to desilt the approaches to the canal mouth. Such scouring sluices, closed with planks carried in grooves, are absolutely essential. Silt is bound to collect above dams of this type and must periodically be removed if the canal intakes and the canals themselves are not to become hopelessly choked and obstructed. All the Moslem dams on the Turia, and most others elsewhere, were equipped with desilting sluices. They were a Moslem development which later Christian Spanish dams were to utilise on a grand scale.

The Mestella dam is typical of the other seven. A particularly interesting feature of all of them is the foundation works.[21] It seems that the mass of masonry in each dam extends something like 15 feet into the river-bed. Below this the whole structure is supported by rows of wooden piles, the tops of which are built into the lowest courses of the masonry. The combined depth of masonry and piles is 20–25 feet. This represents thorough and solid foundation work by any standards, and it is significant that the substructure is so massive compared with the superstructure. It must, however, be remembered that although the Turia's flow for most of the year is only about 400 cusecs, there are occasionally dangerous floods when the flow is more than a hundred times bigger. The dams are then submerged to a depth of nearly twenty feet and must resist the battering of water, stones, rocks and trees. Because they are so low and flat and are provided with deep and very firm foundations, the Turia dams have been able to survive these conditions for a thousand years.

The first four dams supply primary canals which drain ultimately into ravines or rivers to the east and west. Two of these canals feature 'inverted siphons' of considerable size, while on another is a beautiful old aqueduct 750 feet long. The other four dams, the highest being eight feet above the river-bed, all supply canals whose sink is the parent river.

The dams on the Turia may appear to be small, unspectacular and a not particularly notable factor in the history of dam-building; but in fact, for the task they were required to perform and for the conditions under which they had to operate, they turn out to be extremely practical. They continue to meet the irrigation needs of Valencia even today, and it is interesting to note that not only have no more dams been added to the system, but to add at all to it would be pointless anyway. Modern measurements have shown that the eight canals between them have a total capacity slightly less than that of the river. This raises, of course, the question whether or not the Moslems were able to gauge a river and then design their dams and canals to match. At present such a question cannot be answered with any confidence.

The rules governing the operation of Valencia's dams and canals appear to have been instituted early in 961 and the same regulations apply to this day. That they and the works themselves have recently achieved their thousandth anniversary is a remarkable testimony to the quality of both.

North of Valencia the Moslems developed irrigation on the lower reaches of the Rio Mijares, on both banks. Two dams were built, one for each half of the system. At the beginning of the nineteenth century the ruins of one of the dams was still visible, but today nothing remains.[22]

At their southern end the Valencian rice-fields reach to the mouth of the River Jucar. At Antella, the Moslems dammed the Jucar in order to divert water for use in the fields. The dam was built across a sharp bend in the river at a very oblique angle to the direction of flow. Of this original dam nothing has survived,[23] but an inscription over the intake to the main canal suggests that the dam was rebuilt in the thirteenth century; and it was quite definitely rebuilt and enlarged at the end of the eighteenth century by the Duke of Hijar. Even so it is evident that the original tenth-century dam must have been 600 or 700 feet long and, to judge from the level of the canal system, which is substantially original, it must have been about twelve feet high.

According to al-Idrisi, the best paper in the world was made at Jativa, a place of some importance in Moslem times. Two local rivers, the Albaida and Canoles, were the source of Jativa's irrigation, but of the numerous dams which are still in use in the area only three appear to be of Moslem origin; the rest are crude and temporary affairs made of wood and bundles of sticks.[24]

Twenty miles to the east of Jativa at the mouth of the Rio Serpis is another old town, Gandia, home of the Borgias. The Moslem irrigation system of Gandia was based on two dams made of masonry and lime mortar, one on the Serpis and the other on its tributary the Bernisa. A hundred years ago these dams were in a bad state of repair,[25] but this was not crippling to Gandia's irrigation because other dams had been added to the system in A.D. 1500 and then again at the end of the seventeenth century.

Among the Visigothic kingdoms of southern Spain, that of Murcia resisted the Moslem conquest more resolutely than any other; it did not come under complete Moslem domination until 743. The capital city of the province, also called Murcia, was founded on the banks of the Rio Segura about 825 and from the outset this river was vital to the irrigation of one of the hottest and dryest parts of southern Spain. Some of the old dams on the Segura's upper reaches are undoubtedly Moslem or of Moslem origin.

The real interest of the Segura, however, attaches not to its upper reaches but to the last thirty-five miles or so, between Murcia and the sea at Guardamar. Around the city itself the Moslems developed an extensive area of irrigated land, the intention and effect being similar to those at Valencia but on a bigger scale: Murcia's irrigation covers 65 square miles, that at Valencia only 9. The most important difference, though, is that whereas the Valencian scheme used a series of small dams, at Murcia there is only one. This fundamental difference reflects the different characteristics of the rivers themselves. At Valencia, the Turia is flowing in its flood-plain and the surrounding country is nowhere very much higher than the river. Low dams of the type described are all that were required to raise water to the level of the distributing canals. But in the vicinity of Murcia the Segura runs in a deep channel, and a series of dams would have required each structure to be thirty or forty feet high. Such an elaborate and laborious solution was recognised as unnecessary if one properly sited dam was used instead. Such a dam had to be built upriver at a point which would allow, in the first place, a structure of manageable size. Secondly the dam had to be far enough upstream to be at a higher level than the area to be irrigated, so that gravitational flow would guarantee a supply of water. These two factors governing the choice of dam site, namely suitable size and sufficient elevation, had to be weighed against a third consideration: length and route of the supply canals. Clearly the location of the dam could not be considered ideal if it necessitated excessively long canals or canals that had to negotiate obstacles such as deep ravines or high hills. All these considerations, then, had to be balanced, and at Murcia the Moslems produced a thoroughly effective solution (Pl. 25, Figs. 12, 13).

The dam was built just below a sharp bend in the river where the valley is particularly narrow. In recent years parts of it have been rebuilt and altered, so the following description is based on accounts of the nineteenth century when it was in very much its original form.[26]

The main body of the dam was 420 feet long and 25 feet high. For three-quarters of its length it was 160 feet thick at the base, reduced to 125 feet for the other quarter. The two portions of the dam were separated by a low wall running down the face of the structure. The longer of the two sections had a crest two feet lower than the shorter, and it seems clear therefore that the dam was intended to discharge its overflow in two stages. The short section of the crest would only come into action when the long portion was already covered by 2 feet of water.

Considering that the dam was only 25 feet high, base thicknesses of 160 feet and 125 feet seem more than adequate, and as far as stresses and resistance to overturning were concerned so they were. The reason why the dam had such a vast thickness was probably due to the foundation conditions. It is known that although the sides of the valley consist of good strong rock, the bed of the river is very soft and weak, and thus far from ideal for the foundations of a dam. It is supposed that the Moslems built a massive dam, probably carried on a bed of piles, to give the structure enough weight to resist the water-pressure and to ensure that it was not able to slide along the soft river-bed. It is not known, however, what measures, if any, were taken to prevent water percolating through the alluvium and under the dam. That the dam has successfully withstood 1,000 years of use seems to indicate that uplift has not been a problem.

The large surface area of the dam's air face was put to good use. Water flowing over the crest initially fell vertically through a height of 13–17 feet on to a level platform, 26 feet wide, running the length of the dam. This served to dissipate the energy of the water spilling over the crest. The overflow then ran to the foot of the dam over flat or gently sloping sections of the face. In this way the whole dam acted as a spillway

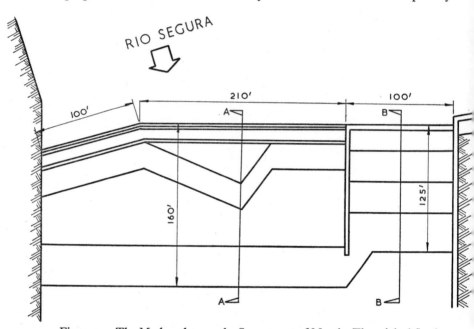

Figure 12   The Moslem dam on the Segura west of Murcia. The original flood-diversion wall extends to the right for 350 ft.

and, moreover, the energy gained by the water in falling 25 feet was dissipated *en route*. Thus the risk of undermining the downstream foundations was greatly reduced. It is for reasons like these that one is inclined to believe that the Moslems had some understanding, albeit empirical, of hydraulics. As usual, rubble masonry and mortar were used for the interior of the dam, and the whole was finished with large masonry blocks. At the right-hand end the dam connects with a long wall, in line with itself and designed to direct flood-waters over the crest so that they could not erode the dam's right-hand abutment. This flood-diversion wall is of interest because, unlike the dam, it has not been recently rebuilt and is believed to be original Moslem work.

The Segura dam feeds two canals, one on each side of the river. These are the main canals which flow more or less parallel to the river along the full length of the irrigated area. They feed many lateral canals and are also equipped with numerous escapes which allow excess water to drain back to the river. As usual there are scouring sluices near the canal intakes. That serving the northern and bigger canal is particularly interesting for two reasons: it is very deep and is closed with a small arched dam. About 500 feet down the canal from the river a deep channel leads off to the right.

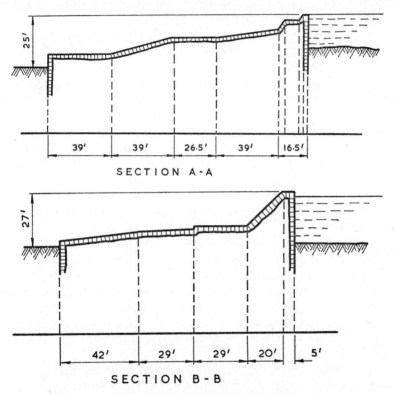

SECTION A-A

SECTION B-B

Figure 13   Cross-sections of the dam shown in Figure 12.

It is clearly artificial and involved an excavation 35 feet wide at the top, 13 feet at the bottom and some 100 feet long. The tiny dam which was built across this cut is $11\frac{1}{2}$ feet thick and curved to a mean radius of 32 feet. It has a height of 25 feet, the same as the main dam on the river, and moreover the crests of the two dams are level with each other. Now it is evident that the scouring sluices, set low down in the small arched dam, are unnecessarily low if they were intended only to desilt the canal; it is only 9 feet or so deep. One is therefore led to the interesting possibility that the scouring channel was also designed to divert the whole river, when necessary, and thereby drain the main dam, perhaps in the interests of maintenance. Indeed it may be that the channel was cut first with the express intention of dewatering the site of the main dam on the Segura when this was about to be built. The layout is undoubtedly of great interest, and the arched dam, despite being small and thick, is characteristic of a style of dam which became typical in Spain. In fact the arched dam may very well not be Moslem work at all but a later addition or alteration. This is not to say, however, that some earlier Moslem device did not achieve one or more of the ends described.

The irrigation of Murcia was based on what might be called the 'one-dam' system. Further down the Segura around the town of Orihuela[27] one finds irrigation based on the 'multi-dam' type of scheme as at Valencia. The first dam, the *azud* of Beniel, is three miles upriver from Orihuela and irrigates some eight square miles of fields; and there is a second dam near by. Much nearer the town two more dams feed a vast network of canals on the northern bank, this network eventually finding its way back to the Segura at points more than fifteen miles from the dams. A century ago one of these dams powered a battery of seven 14-foot-diameter norias which lifted water to another canal. Of the other dams, which are between Orihuela and the sea, three are part of the original system, but the fourth, originally built to power a flour mill, was not incorporated into the irrigation system until around 1600.

Thus the Orihuela region depends to this day for its irrigation on seven Moslem dams plus a converted mill dam which is a later work. The dams follow exactly the pattern of those in Valencia; that is to say they are built of rubble masonry with a cut masonry facing; they are long, low structures with a gently sloping air face over which the river flows. They were built for river diversion only: none of them provides any degree of water storage. In the tenth century the area they irrigated was about fifty square miles.

It can be seen, then, that Moslem irrigation based on the Segura was extensive and thorough. Quite the best way to appreciate the significance of what was achieved is from the hills above Orihuela: as far as one can see, westwards towards Murcia and eastwards towards the Mediterranean, there stretches a broad strip of vegetation, as green and luxuriant as the rest of the landscape is bare and arid.

Twenty-two miles north of Orihuela the old town of Novelda stands on the southern bank of the Rio Vinalapo. The customs which govern irrigation in this area and the distinctive local hydraulic terminology both point to an irrigation system of Moslem origin. But the dam which was built there may be later. Water from various springs near Novelda is collected behind a masonry dam, 700 feet long and $6\frac{1}{2}$ feet thick with deep foundations 14 feet thick.[28] A reservoir dam even of this small size is uncharacteristic of Moslem engineering. So is the fact that the structure is curved, a feature which is more compatible with later Christian dam-building, especially in a region where two big dams were built in the seventeenth century. The origins of the Novelda dam, then, must remain undecided.

About ten miles down the Vinalapo from Novelda is Elche, a town which is full of reminders of 500 years of Moslem occupation, none more noticeable than the plantations of date-palms. Indeed Elche is the only place in Europe where they are grown. The irrigation system developed by the Moslems in Elche to grow dates and other crops is basically the one still in use. The tenth-century dam which supplied water, however, has not survived although it was, so far as can be judged, a typically small diversion dam, a few miles up the river. Most emphatically it was not the present Elche dam, nor was it anything like so big despite the claims, very confused in most cases, of a number of writers.[29] The present Elche dam will be a major topic in the next chapter.

The Moslems' last stronghold in Spain was Granada, a region into which they were driven by the Reconquista in the thirteenth century and from which they were finally ejected 250 years later. During the thirteenth and fourteenth centuries large numbers of refugees, Moslem and Jewish, flocked to Granada, and this increase in population encouraged an extension of the kingdom's irrigation. The already existing works had been built principally in the Vega, a vast, naturally fertile plain west of the city of Granada, watered by the Rio Genil and its tributaries.

Granada itself is situated 2,000 feet above sea level on the northwestern slopes of the Sierra Nevada at the point where the Genil is joined by the Rio Darro. Today it is famous for one thing above all, and nothing typifies Islam's last Spanish fling better than the Alhambra, the 'Red Palace'. It is in connection with this imposing yet exquisite fortress that there are some dams to report. Construction of the Alhambra was begun in A.D. 1248, and from the outset the Rio Darro was used to supply its many fountains, pools, baths and gardens. Where the Darro flows past the steep precipice on top of which the Alhambra is built, it is 400 feet below the palace. It was necessary, therefore, to tap the river well upstream at places where its levels were above that of the Alhambra's gardens.

One of the canals had its intake six miles upriver from Granada, and C. C. Scott-Moncrieff has described[30] the remains of its dam. The structure was built of 'boulder masonry' or what we usually refer to as rubble masonry. It is not known whether it was faced with masonry blocks. The dam had vertical air and water faces and was 12 feet thick and 6 feet high. It was 36 feet long and connected to the sides of the valley by wing walls 100 feet long. The canal's intake was at the northern end of the dam, and just downstream was an escape to allow surplus water back into the river. In 1868 it was reported that 'the original weir was breached, and most of it carried away forty years ago'; little wonder, then, that no trace of it can be found today.

The Darro was also dammed at two other points, and these dams fed canals whose water was used either in the Alhambra or else to drive mills in or near the city.

Above Granada the other local river, the Genil, was the source of supply to the Acequia Real de Genil, 'the Royal Canal of the Genil'. Its dam was built five miles from the city and was already in ruins in the nineteenth century, so much so that a makeshift stake-and-gravel dam was needed to help feed the canal. The original dam was 105 feet long, $11\frac{1}{2}$ feet thick, but only 3 feet high.[31]

The Darro dams and the one on the Genil appear to have been built in the middle of the thirteenth century and were the only ones above Granada. Three other Moslem dams on the Genil, for which there are no details, must have been below the city and to the west, connected with the irrigation of the Vega.

The evidence that the Moslems brought irrigation engineering to Spain is overwhelming and comprises more than the obviously Moslem dam and canal schemes which have been the focus of our attention. There are, for instance, the important crops such as rice, cotton, oranges and even the date-palm, all of whose cultivation in Spain goes back at least 1,000 years. The whole repertoire of eastern water-raising devices—the noria, the saquiyah, the chain of pots, the pot-wheel and the shaduf— were introduced by the Moslems, and substantially original examples of these machines, especially norias, can still be seen in some places. Earlier we noted that in Valencia the 'water laws' pertaining to the irrigation system are uniquely Moslem even today. To varying degrees this is true of other places. Even though the rules and regulations which have been handed down are often not in a written form and have moreover been considerably adjusted and altered at various times, there is general agreement that their origins are tenth century. And finally there is the linguistic factor, namely that Spanish irrigation terminology is almost totally Arabic in origin.

At the same time, however, dam-building in Moslem Spain was limited in its applications in exactly the ways which characterise Moslem work in Persia, Iraq and other parts of the Middle East. Irrigation principally, but also water power and water supply, were the stimuli to the construction of dams, but always river dams, or diversion dams, and never reservoir dams. Should this be a source of surprise or not? In so far as there was no lack of precedent perhaps it should, especially in Spain, because it is at Mérida and Alcantarilla that the three finest of ancient reservoir dams are to be found. It is clear that the Moslems knew these dams because all were given Arabic names.[32] Yet none of these dams was in use in the Middle Ages, nor was the idea copied elsewhere.

Conceivably the Moslems were reluctant to build reservoir dams because it would have involved working on a scale beyond what they regarded as either feasible, economical or safe. It is a basic fact of dam-building that the idea of collecting water when the available flow is in excess of existing needs, and holding it in reserve until the need exceeds the naturally available supply, almost always means building high dams. Nature rarely provides a site at which a low dam will create a large reservoir. It would be inaccurate, however, to give the impression that the Moslems did not practise water storage at all. For public and private water supply they built many underground cisterns and tanks, some very large, and this technique undoubtedly had two advantages: the stored water is kept clean and is less prone to evaporation.

It seems reasonable therefore to conclude that the Moslem engineers' faith in low diversion dams was based entirely on tradition. Their inheritance from Iraq and Persia of large-scale irrigation based on rivers and multiple canal systems was not only highly successful but obviated the need for water storage. The only proviso, of course, is the existence of rivers whose average daily flow is adequate to meet the needs of irrigation. Thus we find that the bulk of Moslem settlement throughout the Islamic Empire was along the banks of big rivers: the pattern is particularly well defined in southern Spain.

5

Christian Spain

THE MOSLEMS' GOLDEN age in Spain ended at the beginning of the eleventh century. Such breakdown of government and general chaos followed the death of the dictator al-Mansur that in 1031 the Umayyad Caliphate was abolished. Moslem Spain's condition continued to weaken, and inevitably the ambitions of the Christian kings in the north began to grow. The progress of the Reconquista took a significant step forward when Toledo was captured in 1085.

In the twelfth century two Berber dynasties, the Almoravids and the Almohades, tried to hold al-Andalus against the Christians. As a result the pace of the reconquest was slowed down for a hundred years or more, but it was never completely halted. And then in 1212 at Las Navas de Tolosa the Almohades suffered a crucial defeat from which they never really recovered. Subsequently Ferdinand III of Castile and Leon and James I of Aragon (James the Conqueror) carried forward the Christian advance with great energy and enthusiasm and almost total success. By the middle of the thirteenth century the Reconquista was all but complete, and yet it was not quite carried through. James turned his attention to conquests in the Mediterranean, Ferdinand died, and the Moslems were thereby allowed to establish themselves for a further two and a half centuries in Granada.

Nevertheless the Christians now occupied the bulk of Moslem Spain, and an inevitable result was their assimilation to varying degrees of Moslem technological skills. Details of this transmission are too complicated for us to discuss at length (nor has the question been thoroughly studied) but a few points should be mentioned.

As the Christians overran southern Spain they necessarily absorbed a large Moslem population. In return for the payment of special taxes, these so-called Mudejars[1] were allowed to retain their own religion and laws; but of more importance to the emergence of a tolerant attitude was the realisation that the economic and material life of the country depended on the expertise of Moslem artisans and workers. Ultimately this tolerant attitude vanished and all Moslems—they had by now become known as Moriscos—were officially expelled from Spain at the beginning of the seventeenth century. But for 200 years or more the Mudejars, among whom were builders, paper and textile makers, manufacturers of iron and leather goods, and expert artists in pottery, ceramics, ivory and precious metals, carried on their traditional skills quite freely. To this extent, then, Moslem Spain lived on under Christian rule. Of most significance for us is the fact that agriculture was almost entirely in the hands of the Mudejars; and thus Moslem irrigation continued to be practised well into the Christian period by those who had developed it and were experts in its management. In fact their skills in agriculture were deemed important enough for some Christian landowners, especially in Valencia and Aragon, to oppose the moves which were eventually made to eject the Moriscos from Spain.

That the great prosperity of al-Andalus depended on dams and canals, water-wheels and irrigated fields was patently obvious to the Christian conquerors. They therefore made efforts to ensure that nothing was done to disrupt their operation, and this was as sensible as it was predictable. Conquest may sometimes require the strategic destruction of public works and utilities, but wilful annihilation of the very basis of the economic life of a captured territory is rare because it is so senseless.

There are many instances of the steps taken by James the Conqueror, and other kings, to preserve Moslem irrigation works and their attendant rules and regulations. One particular example will serve to illustrate what was typical of other places.

On 28 September 1238 James took possession of Valencia along with the irrigation scheme described in Chapter 4. In the following year, 1239, he issued the following edict:[2]

> For us and our successors we give and grant for ever to you all, and to each one of the inhabitants and settlers, of the city and kingdom of Valencia, and of all confines of that kingdom, all and each one of the canals freely and unreservedly, large, medium and small, with their waters, and branches and conduits of water, besides the waters of springs, excepting the royal canal which runs to Puzol, of which canals and springs you shall possess the water and conduits and water courses for ever, always, day and night. And so you may irrigate with them and take the waters without any obligation, service or tribute; and you shall take these waters as was established of old, and was customary in the times of the Saracens.

This document is important for two reasons. It indicates in the first place the nature of the regulations which the Moslems drew up for irrigation works: that the irrigators were, for instance, allowed the use of the canals and the water they carried 'without any obligation, service or tribute'. And secondly, the king's enlightened attitude to the future of the city's irrigation and his intention that the irrigators should be allowed to own and operate the scheme themselves emerges clearly. Even 'the royal canal which runs to Puzol' (it is the first of the eight canals and is called today the Royal Canal of Moncada) was eventually handed over to the people in 1268.

James the Conqueror inherited Moslem irrigation works in areas other than the kingdom of Valencia, notably on the Ebro and its tributaries. These in fact were the first of such specimens that the Aragonese became familiar with, because their conquest of the Ebro valley in general and Saragossa in particular was successfully accomplished before the end of the twelfth century. Details of Moslem dams on the Ebro are lacking, but presumably they were of the usual form. In addition to feeding water to irrigation canals, the dams were also used to make the Ebro navigable and to supply power for mills.

Between Haro and Saragossa there are old dams (and irrigation canals) on tributaries of the Ebro; they are either late Moslem or early Christian work. One is a stone dam at Sartagudo, north-west of Calahorra, which is 160 metres (525 ft) long and 4 metres (13 ft) high; another is a long, low diversion dam near Alagon which has been rebuilt more than once to compensate for its considerable age; and a third, in much better condition, is at Ejea de los Caballeros. At about the same period, dam-based irrigation was also apparently practised on the Rio Alhama, west of Tudela.

With the coming of the thirteenth century we can point with confidence to dams of Christian origin, the first known efforts at dam-building by the Spanish (from now on 'Spanish' will be taken to mean 'Christian'). About 1220 the dam of Cardete[3] was built near Tudela to supply irrigation water to about 750 acres of land on which cereals were grown. The size and shape of the dam are not known.

In 1252 King Theobald I of Navarre gave permission to the people of Fustiñana and Cabanillas, two small towns south-east of Tudela, to take water from the Ebro for irrigation. The river was dammed at a point about two miles east of Tudela, and from there the canal of Tauste was cut for twenty-five miles along the Ebro's north-eastern bank, draining back into the river near Alagon. This canal is still in use and today is fed by a rubble dam 83 metres (270 ft) long and 4 metres (13 ft) high. It is almost certainly not the original dam, but its size and mode of construction are probably similar.

In 1339, on 23 August, Peter IV of Aragon gave the people of Manresa (thirty miles from Barcelona) the rights to utilise the Rio Llobregat.[4] A dam was built fourteen miles upriver from the town to supply thirty miles of canals. Below Manresa, the Llobregat has been used for centuries to drive water-wheels at Papiol and Molins de Rey. These works also point to dams on the river at an early date.

None of the above-mentioned Spanish dams suggests any radical departure from traditional Moslem practice. Irrigation and occasionally water power were provided by low diversion dams across rivers. There is, however, one more dam in the Ebro valley which is of some significance because it represents a new development. When discussing irrigation on the Rio Aguasvivas, Andres Llaurado makes the following comment:[5]

> A certain renown is given to it [the river] by the famous dam or reservoir of Almonacid, 100 *varas* in length by 7 *varas* thick, whose construction is attributed by some authors to the period of Arab domination, and by others to the initiative of James the Conqueror.

If the reservoir of Almonacid is Moslem, then it is the only such specimen in Spain and the only example of a high dam built by the Moslems anywhere. By contrast high dams for water-storage were to become a regular and important feature of Spanish engineering, so that there is every reason to credit the dam to the 'initiative of James the Conqueror'. It is interesting also to note a curious sequence in Spain's dam-building record: reservoir dams were built there by the Romans, but not by the Moslems, and then they reappear in the thirteenth century with the Christians. What prompted their reintroduction is not clear, nor is there anything to suggest that any of James's other dams, on the Ebro and Ter for instance, were anything more than low river dams.

Almonacid de la Cuba (the name means 'Almonacid of the Reservoir') is twenty-five miles south of Saragossa. The town is built on the sides of a deep and rocky gorge through which flows the Rio Aguasvivas. The gorge begins about a quarter of a mile to the south of the town, and here, at a suitably narrow point, the dam was constructed.

Any attempt at an assessment of the dam in its original form is hindered by two factors: both are visible in Plate 26. At some date the dam has been heightened, effectively but crudely. A huge mass of rubble masonry, set in mortar, smoothly faced and vertical on both sides, was grafted on to the original dam along its entire length. As a result it is now impossible to locate the crest of the original dam. Evidently the dam was raised in an effort to offset the effects of the old enemy, silt. In the long run it was unsuccessful, and nowadays the reservoir is full to the brim with soil and under cultivation. Trees growing in the 'reservoir' are clearly visible in Plate 26. Unfortunately, then, it is now quite impossible, short of carrying out a massive clearance of silt, to see anything of the dam's water face. Much about the original dam must be based on conjecture.

A '*vara*' is an old Spanish unit equivalent to 33 inches. Llaurado therefore makes the length of the thirteenth-century dam 275 feet, which

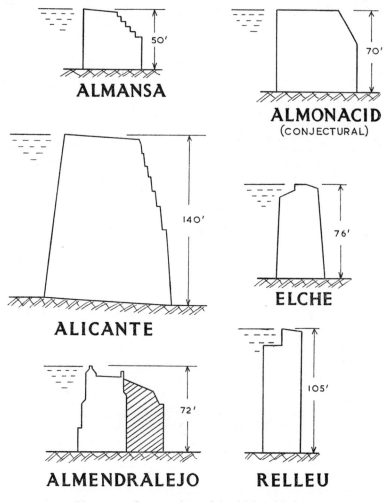

ALMANSA

ALMONACID
(CONJECTURAL)

ALICANTE

ELCHE

ALMENDRALEJO   RELLEU

Figure 14   Cross-sections of six old Spanish dams.

appears to be a reasonable estimate; it may be a little less. The height of the original dam is indeterminate. The extension to the dam in all probability begins part way up the air face (rather than being built up from the crest), so that an unknown amount of the original structure is buried behind the heightening wall. Thus from what is still visible it can be stated that the thirteenth-century dam was certainly not less than 70 feet high; and even this is an impressive figure for the time.

Llaurado gives the dam's thickness as 7 *varas*, nearly 20 feet, but this is surely a mistake. The crest thickness of the raised dam is over 80 feet in places, more than wide enough to carry the road from Almonacid to Belchite. The original structure must have had a comparable crest thickness, and if we conclude that it was approximately as thick as it was high (Fig. 14) we shall be in accord with what a thirteenth-century dam-builder would have regarded as reasonable in terms of stability and safety.

Much of the air face is original work. It consists of large and carefully cut blocks of masonry arranged in a series of steep steps. The joints are filled with a hard and strong mortar. At the left-hand end of the dam much of the original construction was destroyed or removed at some stage. When the dam was raised this portion was rebuilt with a vertical face, and consequently a slice of the old dam's interior has been exposed. It consists of rough rubble masonry embedded in a mixture of stones, earth and lime mortar. Overall, the dam cannot be said to have been well made, which does not mean that it was unsafe: it was, after all, very thick.

Low down in the structure is an outlet tunnel for drawing water from the reservoir. It cannot now be used, and indeed must have been out of action ever since silt first covered its intakes on the water-face side. Seemingly the dam has at no time been required to discharge water over its crest. The sides of the river valley consist of very hard rock, and over-flow has always been directed down a natural rock spillway at the right-hand end. Originally the highest point of this rock face was itself the spillway crest somewhat below the top of the dam. Following the height-ening of the dam, the spillway crest was raised artificially by means of a long curved weir. This weir appears in fact to have been constructed from the cut stone facing taken from the far end of the thirteenth-century dam.

The 'cuba' of Almonacid was built to store irrigation water for land beyond the town, the first instance so far of a reservoir for irrigation. Nature, however, has had the last word. The Aguasvivas, not a very large river, nowadays meanders across the top of the silt-choked reservoir and under normal conditions passes through two tunnels under the spillway crest. Some of the water is used in a mill near the dam and then rejoins the main stream. The whole of the river's flow is then channelled down the side of the gorge and eventually back into the natural river-bed.

Today the dam is both an impressive and strange sight. Its total height is nearly 100 feet; it is very nearly as thick; like some gigantic retaining wall it holds back tons and tons of silt and yet serves as a bridge, 340 feet long, across the gorge. One could almost be excused for not recognising it as a dam at all.

If the dam at Almonacid is significant as marking a return in Spain to the construction of high dams for reservoirs, then the Almansa dam is even more significant. For here is not only a reservoir dam but also an arched dam; it is still fully operational and for several centuries has been cleverly protected from the curse of silt which disabled its predecessor.

The dam stands three miles to the west of Almansa, very near the railway line from Alicante to Madrid. Like the one at Almonacid, the Almansa dam has been heightened, but initially we shall confine our attention to the first stage which is not now incidentally in precisely its original form. It is almost certain that a portion of the original dam's upstream face has at some time been dismantled. The air face, however, is unchanged. The first stage of the Almansa dam was built across a small river at a point where two outcrops of rock reduce the width of the valley to as little as 50 feet. The dam is 48 feet high on the air-face side and perhaps 50 feet high on the water-face side. In other words its crest slopes gently from front to back, and evidently this was intended to facilitate the discharge of overflow, the dam having no separate spillway when first built. Despite the later addition of a wall to heighten the dam, much of the original broad crest is still clearly visible, as Plate 27 shows.

The dam's thickness at its crest must have been about 30 feet and at lower levels it increased to a maximum of some 50 feet. For its height, then, the dam was disproportionately thick even for a gravity dam (Fig. 14). Nevertheless the builders also took the trouble to curve the Almansa dam, the radius at the crest on the air-face side being 26·24 metres (86 ft). Below the crest the air face is stepped through a height of 7·4 metres (24½ ft) and is vertical from there to the ground, a distance of 7·2 metres (23½ ft); the radius to this vertical section is 20 metres (65½ ft). The water face of the dam is vertical throughout. Thus we have at Almansa a dam of constant cross-section and markedly curved in plan—a massive arched dam.

Although not the first arched dam, the Almansa venture was the first such exercise in Spain and marks the beginning of a long and important tradition in that country. Henri Goblot has suggested[6] that the construction of the dam was perhaps influenced by what had already been achieved at the Kebar dam in Persia, whose construction ante-dates the Almansa dam by nearly a century at the very least. Futhermore, as we shall see later in this chapter, the hydraulic arrangements in several old Spanish dams are very similar to those at both Kebar and Saveh. Even so the notion that engineering techniques were transmitted from Mongol Persia to Christian Spain some time after A.D. 1300 is difficult to accept. At an earlier period when the Islamic empire was at its zenith and the means to communicate ideas throughout its length were undoubtedly present, an idea such as M. Goblot's would be more realistic. But in the fourteenth century the Moslem world was more fragmented than it had ever been before, and conditions were hardly conducive to the transmission or diffusion of technology. If there were technical contacts between Persia and Spain at this period, then their nature needs to be demonstrated. Pending some evidence to support what is at present no more than conjecture, it is preferable to assume that the construction of big dams in Persia and Spain proceeded independently.

So as far as Spanish dam-building is concerned, the structure at Almansa was an innovation. It is, of course, easy for the modern engineer to point out that such a massive dam did not need to be arched or, alternatively, that an arched dam did not need to be so thick. Such an observation, however, would be inappropriate. It is simply important to credit the engineers who built the Almansa dam with the appreciation of a correct principle even though they were over-cautious in applying it.

The introduction of the arch principle into Spanish dam-building doubtless reflects an awareness that reservoir dams must above all be safe; the results of a failure did not have to be tested to be appreciated. By adding curvature to a dam whose profile was already substantial enough for a gravity dam, the builders correctly realised that the structure's stability would be increased. Of course 'stability' still remained very much a qualitative concept; the most elementary of structural calculations was still a long way off in the future, while a rational approach to dam design was not evolved until the middle of the nineteenth century. Nevertheless, an understanding of the nature of the problem and the application of intuition and common sense enabled the engineers at Almansa to take a significant step forward. It was not, however, anything like such a big step or such a daring one as the Mongols took at Kebar.

The Almansa dam is built of rubble masonry with a facing of large masonry blocks, the whole work being very solid and exhibiting little trace of deterioration or erosion. At its two ends the structure is firmly keyed into the rock, and nowhere is there any sign of leakage.

At foundation level the dam is equipped with two outlets—tunnels which run right through the dam from face to face. The smaller is 1 metre ($3\frac{1}{4}$ ft) square and connects with a channel to carry irrigation water away to the fields. The outlet is controlled by a bronze sliding gate set in a small chamber on the air-face side of the dam. The second tunnel, situated at the crown of the dam, is a much more interesting affair and a notable development as well. It is 1·3 metres ($4\frac{1}{4}$ ft) wide and 1·5 metres (5 ft) high and was provided to flush deposits of silt from the reservoir. The manner of its operation is apparently effective but patently hazardous.

The mouth of the so-called scouring gallery is kept closed with a series of thick wooden beams held in grooves in the masonry. At intervals of several years workmen go into the tunnel and take the wooden beams away. Usually the silt bed is so thick and compact that nothing budges.

The workmen then retire to the safety of the top of the dam and loosen the silt with long iron bars. No sooner has the silt been pierced than the water takes over and the bulk of the sediment is washed away. This technique is apparently satisfactory but is subject to the drawback that a whole reservoir full of water is discharged along with the silt. Therefore the operation is best carried out in the wet season when it is easier to re-fill the lake.

The principal source of information on the Almansa dam is Maurice Aymard,[7] whose book, *Irrigations du Midi de l'Espagne*, is an essential work for any historian of civil engineering. When Aymard was in Almansa, early in the 1860s, he examined an ancient document setting out details of a meeting which was held in an attempt to resolve certain problems relating to the use of the reservoir and the management of local irrigation. The document was dated 23 January 1586, and the following extract is of particular interest.[8]

> The interested parties say that the work of the reservoir has already cost much and still does, but that this expense is much less important than the benefits afforded the town and its inhabitants, at whose cost the work was constructed and must be finished.

So in 1586 the dam was already in existence and in use, but was still to be completed. The question is, How old was the dam in 1586?

A much-mutilated inscription on the air face bears the date 1*84 and the missing figure is certainly a 3 or a 5; either date would accord with the document of 1586. Several historians[9] have plumped for 1384, in which case the Almansa dam is a very venerable structure. Should 1584 prove to be the correct date then this would fall in a period when Spanish dam-building was at its peak.

Presumably one outcome of the discussions of January 1586 was the completion of the dam. This was achieved very simply, by erecting on the crest of the existing dam a wall 6·1 metres (20 ft) high, 3 metres (10 ft) thick at the top and 3·58 metres (11½ ft) thick at the base. It is in two straight sections, 110 and 170 feet long, intersecting at an angle of 145 degrees near the middle of the dam. Near its right-hand end the wall broadens to a thickness of 35 feet. A plan of the extension superimposed on the original dam is shown in Figure 15. It should be pointed out here that a similar drawing by Aymard[10] is incorrect: his positioning and alignment of the heightening wall are both mistaken although his cross-sectional dimensions are accurate enough. It is curious that this has not been noticed before, and unfortunately a number of writers have been led astray. In particular it has been assumed that the water face of the original dam has always had the same alignment as the heightening wall. It is true that this was the case following the heightening; but it is inconceivable that the original arched dam was anything but a structure of constant profile. What does seem plausible, however, is the notion that the extension of 1586 was built of masonry stripped from the front face of the first dam in such a way that the final structure had, and still has, a water face which is sheer at every point.

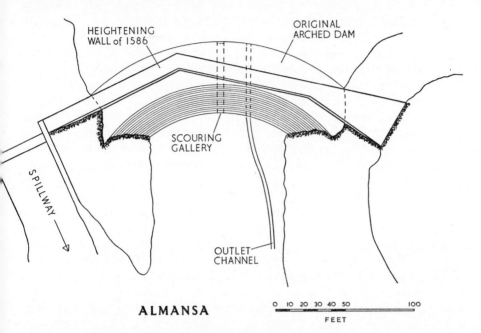

Figure 15   Plan of the dam at Almansa as it is today.

Following the heightening of the dam, overflow was no longer allowed to spill over the crest, but instead a separate side spillway was built at the left-hand end, discharge occurring 6½ feet below crest level. The spillway slopes at a gradient of 1 in 5 and is cut straight into the rock; there is no facing. More recently the spillway crest has been raised a few feet, and in the 1920s an outlet tower was built in the reservoir.

The document of 1586 makes it clear that the cost of building the Almansa dam was borne by the local people. This was also the case at Tibi where a huge dam was built across the Rio Monegre before 1600. This fairly small river, which flows into the Mediterranean a few miles north of Alicante, seems to have been used for irrigation by the Moslems. On 7 August 1579 the idea of building a reservoir on the Monegre was discussed[11] by local irrigators who were keen to extend agriculture around Alicante. Once the proposal was agreed, King Philip II was approached for his approval, which he gave, but he refused a grant from the royal treasury, and instead the promoters of the dam had to be content with his majesty's authorisation to raise the money themselves, apparently by contracting a loan.[12] This and later loans were eventually paid off by increased taxation of the lands around Alicante which had been rendered more productive through the increased irrigation available from the dam.

The construction was begun on 17 August 1580 and in a short time reached a height of seventeen feet. But the funds ran out, enthusiasm waned and the work stopped. For ten years nothing was done. Then in 1590 more money was raised and on October 7 building recommenced. By the beginning of 1592 the dam was fifty feet high, and in July of that year yet more money was raised. Construction proceeded rapidly and by the end of August 1593 the dam was 110 feet high. In October, although still not finished, the dam was put into use and in only three days had impounded 28 feet of water. In 1594 there were more financial setbacks, not helped by the fact that Alicante had to cope with the expense of defending itself against the English fleet. Nevertheless the builders pressed on, and the dam was finished in November 1594. It stood 196 palmos ($134\frac{1}{2}$ ft) high, 4 palmos (33 in) short of what had been planned.[13] All the same a dam $134\frac{1}{2}$ feet high was a tremendous achievement: it was destined to be the highest dam in the world for the best part of three centuries and can still be admired today; it is shown in Plate 28.

The Alicante dam is eleven miles north-west of the town at a point where the Rio Monegre flows through a deep and narrow gorge of hard calcareous rock. The site chosen is eminently suitable for a dam which is big in every sense. Apart from being over 130 feet high, its crest is 20 metres (66 ft) thick and the base 33·7 metres ($110\frac{1}{2}$ ft). Because it was built in a narrow gorge the dam has a maximum crest length of some 80 metres (262 ft) but is only 9 metres (30 ft) long at the base. Nevertheless the total volume of the structure is enormous: more than $1\frac{1}{4}$ million cubic feet constructed, as we have seen, in a total time of about five years.[14]

The Alicante dam, then, is a vast structure even by the standards of gravity dams, but, like its predecessor at Almansa, it is also arched. The radius of the water face at the crest is 107 metres (351 ft), and it is evident that the engineers were not prepared to take any chances at all with the dam's stability. One can hardly blame them. Here, for Spain at least, was a water-retaining structure of unprecedented height, and it is not difficult to appreciate why, in the absence of any form of structural analysis, the engineers elected to build a structure of great thickness to which they added curvature in order to be absolutely safe. That by modern standards it was an expensive and laborious solution is largely irrelevant. The ways in which early engineers solved their structural problems and the costs they incurred need to be judged by contemporary rather than modern standards.

PLATE 1   A general view of the ruins of the Sadd el-Kafara, the oldest dam in the world. (Courtesy of The Royal Geographical Society and Mrs G. W. Murray)

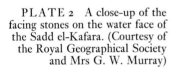

PLATE 2   A close-up of the facing stones on the water face of the Sadd el-Kafara. (Courtesy of the Royal Geographical Society and Mrs G. W. Murray)

PLATE 3   A general view of the remains of the lower of Sennacherib's two dams on the River Khosr. (Courtesy of the Society of Antiquaries of London)

PLATE 4   A close-up of the water face of the lower of Sennacherib's two dams on the River Khosr. The different styles of masonry, suggesting reconstruction, are clearly visible. (Courtesy of the Society of Antiquaries of London)

PLATE 5 The northern end of the Marib dam. On the right are the remains of the 14-metre dam. In the centre are the inlets to the 1,000-metre canal and on the left is the overflow spillway. (From Bowen and Albright: *Archaeological Discoveries in South Arabia*, © 1958 by the American Foundation for the Study of Man)

PLATE 6 The remains of the southern sluices of the Marib dam. The Wadi Dhana, across which the dam was built, runs along the line of trees. (From Bowen and Albright: *Archaeological Discoveries in South Arabia*, © 1958 by the American Foundation for the Study of Man)

PLATE 7　One of the very many Nabataean irrigation and soil-conservation dams which can be found around Ovdat in the Negev. (Courtesy of the Royal Geographical Society)

PLATE 8　This painting which hangs in the Monastery of St Benedict at Subiaco shows St Benedict fishing from the crest of the Roman dam. This is thought to be the earliest surviving picture of a dam. (Courtesy of the Monastery of St Benedict, Subiaco, Italy)

PLATE 9   The remains of a Roman
[dam] in the Wadi Caam, Tripolitania.
[S]mall buttresses are visible and the
[top] of the dam features 'starter blocks'
[rea]dy for an intended heightening of
[the str]ucture. (From *Middle East Science*
[by E]. B. Worthington, reproduced by
[pe]rmission of the Controller of Her
Majesty's Stationery office)

PLATE 10   The Roman dam
at Qasr Khubbaz which once
supplied water to the local frontier
garrison. (Courtesy of the Royal
Geographical Society)

PLATE 11   The Roman dam at Harbaka near Karyatein. It was one of the largest
masonry dams ever built by the Romans and today its reservoir is full of deeply fissured
silt beds.

PLATE 12　The Cornalvo dam near Mérida. Built by the Romans some time in the second century A.D. for water supply, the dam is now used for irrigation.

PLATE 13　The outlet tower of the Cornalvo dam. The springing of the arched bridge, which once connected the tower to the dam's crest, can be seen.

PLATE 14   The Proserpina dam near Mérida. This picture shows the water-face wall with its exposed concrete core and masonry buttresses. The top of an outlet well is visible above the grass-covered embankment.

PLATE 15   The Aivat Bendi, one of the Byzantine dams built to supply water to Constantinople. (Courtesy of the American Society of Civil Engineers and Professor Waldo E. Smith)

PLATE 16   This heavily silted dam near Kurnub in the Negev dates from Byzantine times. (Courtesy of Israel Information Services, New York)

PLATE 17   The Band-i-Mizan, a bridge-dam of the third century A.D. which winds its way across the River Karun at Shushtar. The pointed arches are an indication of later reconstructions.

PLATE 18   Part of the fourth-century bridge-dam on the Ab-i-Diz at Dizful. (Reprinted from *The Traditional Crafts of Persia* by H. E. Wulff, by permission of the M.I.T. Press, Cambridge, Massachusetts. Copyright © 1966 by the Massachusetts Institute of Technology)

PLATE 19   The Saveh dam in Persia. Because of pervious foundations the dam has never held water but has served for hundreds of years as a bridge across the Qara Chai river.

PLATE 20   The Keba dam near Qum. Despite being heightened in Safavid times the reservoir is today full to the brim with silt. Nevertheless this is the earliest authenticated example of an arch dam. (Courtesy of H. Goblot and the Centre International du Synthèse)

PLATE 21   The Pul-i-Khadju, the ultimate achievement in Persian bridge-dams. Built in the seventeenth century by Shah Abbas II, it is still one of the sights of Isfahan. (Photographed by Gordon Porter and Brian Large)

PLATE 22   The 1,000-year-old Band-i-Amir on the River Kur. This is a view of the air-face side and at the left can be seen part of the outlet sluices. (Reprinted from *The Traditional Crafts of Persia* by H. E. Wulff, by permission of the M.I.T. Press, Cambridge, Massachusetts. Copyright © 1966 by the Massachusetts Institute of Technology)

PLATE 23   The zig-zag Moslem dam across the Guadalquivir at Cordoba was built to power water mills—a ruined mill house is visible—and a noria.

PLATE 24   The *azud* of Mestella, one of the eight small river dams which continues to irrigate Valencia and its environs after 1,000 years of use.

PLATE 25   The much rebuilt Moslem dam which raises the level of the Rio Segura in order to irrigate a large area around Murcia in southern Spain. The farthest portion of the dam is believed to be original tenth-century work.

PLATE 26   The dam of Almonacid de la Cuba. The lower stepped portion is original while the upper section of lighter coloured masonry is a later heightening. At the right-hand end, carrying a road bridge, is the spillway, and beyond the dam trees are growing in the silt-choked reservoir.

PLATE 27   The curved and stepped section of masonry is the air face of the four-teenth-century Almansa dam. On to its broad crest a heightening wall was constructed at the end of the sixteenth century.

PLATE 28 The magnificent Alicante dam which was completed in 1594 and was the highest in the world for the better part of 300 years. The original outlet tunnel is just visible behind its replacement of 1943.

PLATE 29 The first arch dam built in Spain is near Elche. The fact that overflow occurs at one end suggests settlement of the far abutment. The top of the outlet well projects into the reservoir in this thin curved dam.

PLATE 30 A sheet of water hides the fact that the Relleu dam is heavily silted. Built in the seventeenth century and raised in 1879, the dam is hardly used today. The opening half-way down the air face provides access to the valves which are fed from the water-face outlet well.

The dam of Ansotegui near Marquina.   The dam of Osiyan near Marquina.

PLATE 31 Four of the buttress dams built by Don Pedro Bernardo Villarreal
de Berriz in Vizcaya, N. Spain, in the first half of the eighteenth century.

The dam of Guizaburuaga near the village
of the same name.

The dam of Bedia on the Rio Ibaizabal.

PLATE 32   The dam of Almendralejo or Albuera de Feria. Built in 1747, it is the earliest buttress dam and one of the first to contain a hydro-power device within the body of the structure.

PLATE 33   The first and largest of the dams which supplied water to the ore-crushing mills at Potosí in Bolivia. The Chalviri dam was built at the end of the sixteenth century and renovated in the 1930s. (Reprinted from the *Geographical Review*, Vol. 26, 1936, copyrighted by the American Geographical Society of New York)

PLATE 34 The eighteenth-century Pabellon dam near Aguascalientes, C. Mexico. The untidy appearance of the structure is due to the fact that the work of heightening it was never properly finished. In the foreground are the remains of the water-powered flour mill. (Courtesy of *Engineering News-Record* and Mr Julian Hinds)

PLATE 35 The Presa de los Arcos is a contemporary of the Pabellon dam (Plate 34). Unfinished buttresses and 'starter blocks' all along the crest indicate that in this structure also a heightening was never completed. (Courtesy of *Engineering News-Record* and Mr Julian Hinds)

PLATE 36 The San Raphael dam in the state of Hidalgo is equipped with two massive desilting galleries discharging on to a masonry apron. (Courtesy of the American Society of Civil Engineers)

PLATE 37 Dating from the middle of the eighteenth century, the Espada diversion dam diverts irrigation water from the San Antonio river. It is one of the oldest dams in the United States. (Courtesy of the American Society of Civil Engineers)

PLATE 38 The remains of the Old Mission dam as they appeared at the beginning of the twentieth century. The portion shown is the centre of the dam with the sluice opening. (Courtesy of John Wiley and Sons)

PLATE 39 The rebuilt and refaced Cento dam on the River Savio near Cesena dates from about 1450. The mill house is at the far end and its tail-race tunnel discharges near the air face of the dam.

PLATE 40 The longest section of the strange Z-shaped dam of the Lago di Terna-vasso near Turin. The dam was built about 1600 and in spite of being heavily silted it is still possible to draw water from the reservoir through the outlet well shown.

PLATE 41   The Ponte Alto
dam near Trento was the first arch
dam in Europe. Begun in 1611,
successive heightenings brought
the structure to its present
height of 124 feet in 1887. It
is one of the most spectacular
of all old dams.

PLATE 42   The Lampy
dam was built between 1777
and 1781 to augment the
Canal du Midi's water supply.
It was the second European
buttress dam following that
of Almendralejo by thirty
years.

PLATE 43   The dam of St-Ferréol was built between 1667 and 1671 and still supplies water to the celebrated Canal du Midi. It was the greatest French dam-building achievement of the seventeenth century.

PLATE 44   The little-known Caromb dam dates from 1766. Originally used to feed water to the canal of Carpentras, the structure is now virtually out of service.

PLATE 45 Nearly 800 years has been ample time for the earth dam at Alresford to become covered in trees and bushes. Nowadays the reservoir helps to supply the water-cress beds seen in the foreground.

PLATE 46 This weir of 1676 stands close to the site of 'Trew's weir', the latter having been built in 1566 to regulate the famous Exeter canal. The present dam performs the same function and also supplies water to the mills on the far bank of the River Exe.

PLATE 47   John Smeaton's dam on the River Coquet was built nearly 200 years ago to feed water to a nearby ironworks.

PLATE 48   The highest of all eighteenth-century English canal dams is Todd Brook near Whaley Bridge. It supplies water to the Peak Forest canal.

PLATE 49 An aerial view of Glencorse reservoir, a part of Edinburgh's water-supply system. Thomas Telford's earth dam of 1823 is the embankment in the lower left corner. (Courtesy of the South-East of Scotland Water Board)

PLATE 50 The Entwistle dam near Bolton was the first in England to top 100 feet. Originally built to power mills, it is now used for water supply.

PLATE 53 The celebrated Zola dam near Aix-en-Provence. t was built by François Zola, father of Emile, and was the first arch dam in France.

PLATE 54 The first buttress dam of the multiple-arch type. The Meer Allum dam near Hyderabad was built about 1800, but details of its origins are otherwise a mystery. (Courtesy of the American Society of Civil Engineers)

PLATE 55   The Lake Vyrnwy dam was the first large masonry dam in Great Britain. Completed in 1892, it formed, at the time, the largest reservoir in Europe. Water from the reservoir is piped nearly seventy miles to Liverpool. (Courtesy of the City Engineer's Department, Liverpool)

PLATE 56   The Sweetwater dam in California at the height of the colossal flood of 1895 which poured over the dam for two days, an experience which the structure survived. (Courtesy of John Wiley and Sons)

PLATE 57   A contemporary photograph of the breach in the Dale Dyke dam near Sheffield soon after the failure of 11 March 1864. This remains the most serious dam failure to have occurred in Great Britain.

PLATE 58   The magnificent Hoover dam is one of the great monuments of dam engineering. It stands to a height of more than 700 feet across the Colorado river. (Courtesy of the Bureau of Reclamation, U.S. Department of the Interior)

PLATE 59   The Grand Coulee dam was the first man-made structure to exceed in volume the Great Pyramid of Cheops. The Franklin D. Roosevelt Lake reaches 150 miles north to the Canadian border. (Courtesy of the Bureau of Reclamation, U.S. Department of the Interior)

PLATE 60   The unique Coolidge dam near Globe, Arizona, is 250 feet high and consists of three huge domes of reinforced concrete. (Courtesy of the Bureau of Indian Affairs, U.S. Department of the Interior)

PLATE 61　The huge sweep of the Mangla dam in Pakistan. This is the largest dam
three which, between them, impound one of the world's largest reservoirs. (Courtesy
Binnie & Partners)

PLATE 62　Lake Trawsfynnyd in North Wales is formed by
four dams. The one shown is the largest and was the first arch
dam ever built in Great Britain. It is 96 feet high.

PLATE 63　A downstream view of Pitlochry dam and power
tion on the River Tummel in Scotland. The fish pass is on the
t. (Courtesy of the North of Scotland Hydro-Electric Board)

PLATE 64   The Kariba dam seen here with all its outlets wide open. The reservoir formed, Lake Kariba, is the second largest man-made lake in the world. (Courtesy of the Director, Zambia Information Services)

PLATE 65   One of the structur[] models used in the design of the Tang-e Soleyman dam. The pict[] shows the model after a test to ru[] (Courtesy of the Institution of Ci[] Engineers, *Proceedings*, July 1962 Paper No. 6600 and Sir Alexand[] Gibb and Partners)

Typically, the dam is built of rubble masonry set in mortar and faced all over with masonry blocks 1½ feet wide by some 2 feet in length. Neither face is vertical, but while the air face is stepped—the thickness increasing by 2 or 3 feet at intervals of between 9 and 17 feet—the water face is smooth (Fig. 14).

A neat device is used to draw water from the reservoir. Close to the water face a vertical shaft 2½ feet in diameter was built into the dam extending the full depth of the structure. This shaft connects with the reservoir by means of 51 pairs of holes, each one measuring 4½ by 9 inches and the highest pair being 23 feet below crest level. Thus water can flow into the outlet shaft even when the reservoir is heavily silted. At its base, the shaft connects with a horizontal gallery, 2 feet wide and 5½ feet high, which runs through the base of the dam to a bronze sliding sluice gate operated by gears and a toothed rack.

There is also a scouring gallery, in principle the same as the one at Almansa but bigger. Its mouth is 6 feet wide by 9 feet high, and these dimensions increase to 13 and 19 feet respectively at the air face. The 'taper' of the scouring gallery is most interesting because the only plausible explanation is that it is designed to make use of the principle of the diffuser or draft tube.

On 13 November 1697 the Alicante dam was badly damaged, apparently as a result of a heavy flood pouring over the crest and cutting deep into the masonry.[15] The dam's usefulness was much reduced until 1738 when the damage was repaired. It was then that the dam's crest was given the slight inclination which it still has. Clearly this was provided to aid the discharge of overflow, and even the later addition of a separate spillway has not altered this practice; the spillway has rarely been used and today is permanently closed with massive wooden beams.

In the very year that work was restarted on the Alicante dam (1590), moves were afoot to undertake the construction of a large dam at Elche. The old Moslem irrigation canals were still of crucial importance to the town's prosperity even after 600 years, but the available supply of water was no longer sufficient: something more than a diversion dam on the Vinalapo was required. Plans for a reservoir dam were drawn up, but nothing was done immediately. Not until 1642 did work begin, and there is no indication in this case how long construction was in progress.

The Elche dam is three miles to the north-west of the town on or near the site of the Moslem dam. At this point a large shoulder of rock outcrops in the river valley towards the left-hand side. The main body of the dam was erected between this rock and the right-hand bank. At the left-hand end of the main body of the dam a short curved wall plugs a gap in the rock, while at the other end there is a straight wall 35 metres (115 ft)

E

long running upstream at right angles to the dam. Plate 29 should make the layout clear.

The dam has a total height of 23·2 metres (76 ft); it is 9 metres (29½ ft) thick at the crest and 12 metres (39 ft) at the base (Fig. 14). These dimensions do not represent a particularly thin dam—the height to mean thickness ratio is rather more than 2—but by a small margin the profile is too slender for a gravity dam. Evidently the builders realised this, and the dam is therefore curved to a mean radius of 62·6 metres (205 ft) and through an angle of 60 degrees at the crest. It was the first arch dam to be built in Spain.

The crest slopes towards both faces, although its upstream portion is obscured when the reservoir is full because of a 3-foot step at the edge of the flat top. The full length of the crest was designed to discharge overflow, but in 1793 the dam was severely damaged by a flood. Ibarra y Ruiz has claimed[16] that the damage was so extensive that a completely new dam was required, and consequently he attributes the present structure to the early nineteenth century. That this is not so is evident from Antonio Cavanilles' description of 1797[17] which corresponds exactly to the dam as it is now.

The damage of 1793 must have been repaired because in 1836 floods struck again, the masonry facing blocks were washed away, and the water soon cut a great hole in the rubble masonry core. The dam was repaired in 1841–42, but even so no proper spillway was built either then or since. Overflow still passes over the top of the dam but normally only at the left-hand end, there being a difference in level along the crest of some eighteen inches. It is just possible that this has been arranged deliberately so that the fall of water will be broken on the outcrop of rock; but it is more likely that the dam has settled at one end.

Because the Elche dam is so much thinner than that of Alicante, water is drawn from the reservoir through a vertical outlet well built on to the dam on the water-face side. The well connects with the lake by means of a series of openings at regular intervals as at Alicante. Originally the outflow was controlled by a bronze gate located at the downstream end of a horizontal tunnel in the base of the dam. This has since been replaced by valves operated from the top of the outlet well at crest level. There is also a scouring gallery just like the previous ones in principle, but with one refinement: a small chamber was built above the scouring tunnel, and this enables workmen to remove the timber beams in relative safety.

In his valuable study of the history of Spanish engineering, Pablo de Alzola y Minondo refers[18] to the 'reservoir of Bellen' built in the province of Alicante during the reign of Philip IV (1621–65). There is little doubt that this is the reservoir which is still to be found on the Rio Amadorio, nine miles west of Benidorm near the village of Relleu. The Relleu dam (Pl. 30), a contemporary of that at Elche, is built in a deep and very narrow gorge of hard limestone rock; the site is ideal for an arch dam and such a structure was chosen. In its original form it stood to a height of 28 metres (92 ft) with a uniform thickness of 10 metres (33 ft) and a mean radius of curvature of 65 metres (213 ft). The crest length was 80 feet. Half-way down, the gorge narrows suddenly to a width of 50 feet, and from this level the dam tapers into the depths of the gorge until it is a mere 6 feet wide at the bottom. The site was splendidly chosen, and it says much for the engineer's understanding of dam-building that he constructed an arch dam whose height/thickness ratio is very nearly 3.[19] Such a figure had not been achieved in Spain before.

The Relleu dam was equipped with a very large scouring gallery set deep in the ravine but it appears that its operation has not been altogether successful. Silt accumulated in the reservoir to such an extent that in 1879 the dam had to be raised. On the crest of the dam and flush with the air face a wall was built, 3·85 metres (12½ ft) high and 5 metres (17 ft) thick. A large rectangular extension of masonry projecting into the reservoir contains the extended outlet well which in the original dam is a circular shaft built into the dam's water face. Water flows into the outlet well through a series of openings and the outflow is controlled by means of a sluice above the scouring gallery.

The Relleu dam is not in good condition and the air face is decidedly wet from leakage through the joints between the masonry facing blocks. There is no spillway, and overflow passing across the crest has removed many of the facing blocks including several on the air face at one end. This has exposed the usual rubble masonry and mortar core. It is regrettable that such a fine example of dam-building is in such a shabby condition, and it is to be hoped that the recent construction of a large reservoir further down the Amadorio will not result in its further neglect.

The dams of Almansa, Alicante, Elche and Relleu were all built to store water for irrigation. They can be regarded as a logical development of the existing concepts of irrigation due to the Moslems. But whereas the tenth-century schemes relied on a supply that was subject to the uncontrollable variations in a river's day-to-day flow, later Spanish engineers saw how to provide a supply that was both easier to regulate and made use of more of a river's total discharge. In other words rivers were not merely being utilised, they were also being controlled and harnessed. What is rather surprising, however, is that the largest of the four reservoirs (Alicante) is comparable in capacity only with the *smaller* of the two Roman ones at Mérida—Proserpina.[20]

A defect in all the dams as originally built is the lack of spillways, and even the two which were added have rarely, if ever, been used. It says much for the quality of the construction of all the dams that they have, with occasional repairs, withstood the effects of what is fundamentally bad practice. By contrast the other hydraulic arrangements are a notable advance. All but the Almansa dam can be drawn down even in the presence of a considerable amount of silt, while the means of clearing the reservoirs of silt are effective enough to have allowed uninterrupted use for three centuries and more. Without scouring galleries the reservoirs would definitely have been choked in a very short time.

We can safely assume that all four dams were built in the same way, one of the first stages being the completion of the desilting galleries and outlet tunnels. Presumably these openings were then immediately used to pass the flow of the rivers which were ultimately to feed the reservoirs. Certainly there is nothing to suggest that the rivers were temporarily diverted. The foundation grooves in the valley sides having already been excavated, the dams were then built up from the base. It is probable that the facing was raised first, and rubble masonry and mortar filling poured into the space formed.

The four dams are an interesting example of what one might call 'structural evolution': that is to say, the tendency to develop the theoretical ideal even in the absence of analytical methods of design. Following the cautious first steps at Almansa and Alicante, confidence in the idea of arch dams grew quickly and produced the slender and elegant structures of Elche and Relleu. Indeed, the Elche dam is a particularly daring piece of work whose length/height ratio of 3 is very high for such an early venture.

Considering the importance of the structural and hydraulic concepts involved and the fine structures which resulted, it is a pity that we know almost nothing of the engineers responsible. There is some information for the Alicante dam. When Philip II first heard of the scheme he sent Juanelo Turriano to examine its feasibility.[21] Turriano was the Italian clockmaker who became court mathematician to the Emperor Charles V and later, as engineer to Philip II, built the famous Toledo water-works. His interest in hydraulic engineering was considerable, as his Codex of 1569 indicates.[22] However, although it contains some discussion of river dams and canal systems on the old-established Moslem pattern, it has nothing to say about big dams, and of course nothing on the Alicante project which post-dates the Codex by twenty years.

Juanelo Turriano reported favourably on the idea of a reservoir on the Monegre and prepared a plan on the basis of which work was begun. The nature of his plan is uncertain; he may have suggested a design for the dam itself or perhaps merely the general arrangement of the scheme. It is believed by some writers[23] that another well-known figure of the time was in some way connected with the work—the architect Juan de

Herrera, designer of the Escorial Palace. Philip II was involved with the Alicante project throughout, and as royal architect Herrera was perhaps called in to advise. Moreover, and the king certainly did send at least one engineer to the site; he was Cristobal Antonelli who took over as chief engineer in October 1590 at a salary of forty ducats a month. For the other dams we know nothing but the name of the engineer of the Elche dam—Johannes del Temple.[24]

Whatever Juan de Herrera may or may not have contributed to the Alicante dam, he did build one dam for Philip II, namely the dam of Ontigola.[25] A straight masonry wall 280 metres (920 ft) long and 7 metres (23 ft) thick, it was built to irrigate the gardens of the summer palace at Aranjuez. In the seventeenth century there are other examples of straight masonry gravity dams. The dam of Pulgar, for instance, was built near Tudela in 1628: it is 110 metres (360 ft) long and 8 metres (26 ft) high. The dam of Elda, sixteen miles north-west of Elche, was the second dam on the Vinalapo.[26] It is 38 feet high, $27\frac{1}{2}$ feet thick at the crest and over 300 feet long. Another dam whose inception dates from the reign of Philip IV is near Huesca on the Rio Isuela. It was planned in 1656 but took thirty years to materialise, and it was not until 1704 that the work was finished. Originally it was 35 metres (115 ft) long, 12 metres (40 ft) thick and 20 metres (66 ft) high, but it has recently been rebuilt and heightened.[27]

The significant point about all these structures is that they are *gravity* dams. In the seventeenth century Spanish civil engineers had reached the point where they adapted the form of their dams to the topography of the sites. Arch dams were built in narrow emplacements where the foundations were good hard rock; gravity dams were used at sites which were wide and shallow.

Late in the sixteenth century and during the seventeenth, Spanish dam-building was in a class all of its own and much superior to anything which had been produced before. In terms of numbers, size and hydraulic and structural concepts, Spanish dams were without equal or rival, in Europe or anywhere else. The works we have discussed reflect the expertise of Spanish engineers in Christian Spain's golden age and even into its sad decline. It was to be a century or more before other European countries produced anything comparable and in comparable numbers. This was not owing to any inherent lack of engineering skill; rather it was the lack of a need. The reasons for Spain's early emergence as a dam-building nation are essentially three: the need for water in a country whose climate is semi-arid; the long-established Moslem techniques of irrigation based on practical and equitable rules and regulations; and engineering know-how.

In 1736 a book called *Maquinas Hydraulicas de Molinos y Herrerias, y*

Govierno de los Arboles y Montes de Vizcaya was published in Madrid.[28] Its author was a certain Don Pedro Bernardo Villarreal de Berriz, a knight of the Order of Santiago and a Basque nobleman. He was also the owner of mills and forges in the province of Vizcaya, and these together with the lands he owned kept him fully occupied throughout his life.

Mills and forges require supplies of water, and in the wet mountains of Vizcaya these were readily available from local rivers across which low diversion dams were built. Don Pedro's book contains sections on dams, and his discussion is of interest on two accounts. Firstly, he was one of the first engineers to write on the subject of dams and particularly to offer advice and instruction on their design. In this context his book will be discussed in Chapter 9. Secondly, based on his 'theories' of dams, he introduced some new ideas which can still be seen in a number of surviving structures.

In Don Pedro's time the traditional river dam was a long straight gravity structure made of masonry. Don Pedro addressed himself to the idea that improvements would result from the use of the arch form instead. He seems to have sensed also that for a given height of dam there was a limit to the span over which an arch would be effective and safe. Thus he writes, 'When the river bed is narrow, one arch is sufficient, and when it is wide, two, three, four or five will be needed.'[29]

Furthermore, Don Pedro saw that whereas a single arch could be fully supported at its natural abutments, the dam with two arches or more would require support from artificial abutments. Thus his multiple-arch dams feature heavy masonry buttresses at the ends of the intermediate arches.

The arch form in Don Pedro's dams is not fully developed. The air faces of his dams are curved between the buttresses, but, with one exception, not the water faces. On the upstream side the structures are straight throughout their full length. Hence the best way to visualise the arrangement is to imagine a masonry arch bridge lying on its side with the deck facing the water; Plate 31 shows the typical layout. As A. del Aguila has pointed out, Don Pedro was a pioneer in the use of buttress dams although he did not himself see things in quite that way. Basically he was using the arch principle in special circumstances and produced a type of buttress dam as a result.

In his book Don Pedro Bernardo Villarreal de Berriz describes five dams built to his specifications. Two are near Marquina on a tributary of the Rio Artibay. The dam of Ansotegui (Pl. 31) is a two-part structure, an arch dam at one end and a buttress dam at the other. The arch spans the river itself, the right bank as one abutment and a huge masonry buttress as the other. The arch is 20 feet high, 50 feet long and of varying thickness, 6 feet at each end and 3 feet at the crown.

Beyond the big masonry buttress the rest of the dam comprises a 450-foot-long masonry wall built on what originally was low-lying land beside the river. Its height ranges from 12 feet at the end adjacent to the arch, to nothing at the other. A selection of masonry buttresses, 5 feet wide and 6 feet long, support the wall at various arbitrarily chosen points.

The dam of Ansotegui still feeds water into a small canal on the right bank, but the small reservoir it once formed is nearly full to the top with silt and under cultivation. The dam is without doubt the least representative of Don Pedro's notions on diversion dams, but the other structure near Marquina, the dam of Osiyan (Pl. 31), is a perfect example. The straight water face is 75 feet long and its two arches are supported by masonry walls at each end and a heavy buttress 7 feet wide, 9 feet long and 10 feet high in the middle of the river. The arches are 7 feet thick at each end and 3 feet at their crowns. Nothing survives to suggest where the mill or forge once stood, but the disused sluice-gates are still visible.

The dam of Laisota is near Guizaburuaga, about seven miles east of Guernica. It is a replica of the twin-arch dam near Marquina but larger. The straight water-face wall is 110 feet long and 14 feet high. The two arches each span 40 feet and have a crown thickness of 6 feet. The central buttress is 8 feet wide and 7 feet long, so that the dam's cross-section at the centre is 14 feet high by 20 feet thick. At the right-hand end a steep stepped spillway takes care of excess overflow (all normal overflow pours over the crest), but there is now no trace of any canals to mills, forges or anything else. At some time a sloping masonry wall has been added to the dam's upstream face, presumably to counteract leakage. This is the only one of Don Pedro's dams to feature a low-level outlet. It is at the crown of one of the arches, but its purpose is not clear. It could conceivably be a scouring sluice or perhaps a gallery to drain the space above the dam.

Near the Laisota dam are two more old buttress dams; one is 30 feet high with a straight wall and a single massive buttress. It is obviously not one of Don Pedro's and is probably much later.

Two kilometres downriver from Guizaburuaga is a three-arch dam which Don Pedro describes thus: 'And I have built at Guizaburuaga another of the same type, with three arches of unequal length in order to utilise some large projecting rocks.'[30]

In order to locate the buttresses firmly on these rocks, he made the central arch 35 feet long and the others 25 and 30 feet respectively. The buttresses are each 8 feet wide, 6 feet long and 5 feet high, the same height as the dam. The dam has a straight water-face wall 130 feet long. A curious feature is that the longest arch is the thinnest at its crown (3 feet) and the shortest is thickest (6 feet). At some point the dam has been heightened 15 inches by means of a narrow parapet wall. The dam is shown in Plate 31.

The largest of all the dams mentioned by Don Pedro is at Bedia on the Rio Ibaizabal, a tributary of the Rio Nervion. It would seem that a single-arch dam 53 metres (175 ft) long on the same site proved to be in-adequate and was replaced by a five-arch structure of the Don Pedro type. Today, however, the dam is not arched, which indicates a later reconstruc-tion. As it is now (Pl. 31) the 175-foot-long dam is essentially a straight masonry wall 12 feet high and about 10 feet thick backed up with four large buttresses 20 feet long and 15 feet wide. Even in the summer the dam discharges a tremendous volume of water over its crest and also diverts several dozen cusecs into a canal on the northern bank; the dam is fully employed by a local mill.

Contemporary with the works of Don Pedro Bernardo Villarreal de Berriz in northern Spain is the prototype of all big modern buttress dams (Pl. 32). It still stands near Almendralejo and thirty-two miles south of Badajoz. Old documents[31] state that it was built in 1747 to power a flour-mill which served the local villages of Feria and La Parra. The dam is shown in Figure 14.

An examination of the dam suggests that during its life there have been several modifications and alterations. Originally it must have been about 64 feet high, but subsequently it was raised to 72 feet and recently, since 1936, to 77 feet. Its crest length is now 400 feet. When first built the dam had a crest thickness of 32 feet and a base thickness of 40 feet, and was just heavy enough to perform as a gravity dam. But the engineer, in order to be absolutely sure, added buttresses to the air face. There are five big ones near the middle and two smaller ones at the sides. They are all 3·2 metres (10½ ft) wide; the five big ones are 38 feet long and the remaining two, 26 feet long. The top corner of each buttress appears to meet the dam at the level of its original crest, and from there each buttress slopes down-wards at an angle of 25 degrees to the horizontal. Inspection of the but-tresses seems to indicate that, when the dam was raised the first time, the five central buttresses were lengthened by 12 feet; hence the difference in length between them and those at each end.

The spaces between the five big buttresses are enclosed. A thick wall runs across the ends of the buttresses, each enclosure is vaulted over with an arch, and above each vault is a plane sloping roof. One of these en-closures, the second from the left, once housed the flour mill and its water-wheel. Very little of this device has survived, although it is evident that the wheel was fed by a low-level pipe running through the wall of the dam. Nowadays this is blocked off.

The Almendralejo dam (also called the dam of Albuera de Feria) is one of the earliest examples of a big dam built to *store* water for a hydro-power unit, elementary though this machine must have been. What is more, it is one of the first dams to have contained a water-power device actually within the body of the structure. It is also worth mentioning what must certainly be a unique feature: at some time one of the inter-buttress enclosures was used as a chapel.

The dam is made of rubble masonry throughout. There is no special facing, although the exterior of the dam has been worked to a smooth finish. Founded on rock, the structure is extremely well made and still in excellent condition, with no sign whatsoever of leakage.

Originally overflow discharged over the dam's crest, and this means that the first corn mill was not a part of the dam. It may well have been among the ruins of mills which have been found downstream. When the buttresses were roofed over to house the mill, overflow was diverted to a special side spillway, and following the first heightening a second spill- way was added. Both are still used.

The dam has long since ceased to be used for milling and today it supplies water to Almendralejo. Nevertheless it is a structure of the first importance, and a pioneer in that it was the first big dam in which the buttressing technique was utilised in the modern fashion.

A great deal has been written on Spain's decline, which was detectable at the end of the sixteenth century, advanced rapidly during the seventeenth and was widespread and deep-rooted by the eighteenth century. Many reasons for this decline have been put forward: frequent debasements of the coinage; too many civil servants and too few workers; the stranglehold which the Church had on so many aspects of Spanish life; a failure or perhaps an inability to develop science; the expulsion of the Jews and Moslems. In fact it is difficult to decide to what extent these elements were causes or effects. Perhaps the onset of decline in any country is impossible to explain completely or simply; certainly this is true in the case of Spain.

Spanish agriculture declined steadily along with everything else. Some authorities have argued that Morisco expulsion early in the seventeenth century was a principal cause. Another contributory factor is said to have been the steady growth of sheep-rearing at the expense of arable farming. Driving millions of merinos all over the country must have acted to the detriment of much cultivated land; it is also symptomatic of the fact that the only labour force available was small, unskilled and cheap to hire.

The downward trend in Spain's agricultural activities is reflected in the dam-building record. Following the fine crop of dams built around 1600, there is subsequently increasingly little to report. The eighteenth-century dams of Vizcaya and Almendralejo, although significant tech- nically, were power dams and not for irrigation. Not until the reign of Charles III (1759–88) did a Spanish ruler attempt to revitalise the country's economy through financial reforms, aids to commerce and encouragement of agriculture. As part of the latter, irrigation canals such as those of Huescar and Manzaneres were built, and additions were made to the existing canal systems on the Ebro, Segre and Jucar.

Towards the end of Charles III's reign, three dam-building projects were under way, the biggest that had ever been undertaken even in Spain.

The Rio Guadalentin is a tributary of the Segura, and north-west of Lorca two large dams were begun on this river in 1785 and finished six years later. Ultimately both had to be replaced, but nevertheless a good deal is known about them from contemporary descriptions.[32]

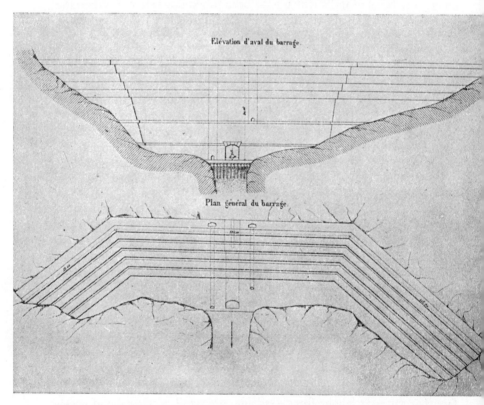

Elévation d'aval du barrage.

Plan général du barrage.

Figure 16 Maurice Aymard's plan and elevation of the Puentes dam. The dam collapsed in 1802 when the pressure of 154 ft of water blew out the piled foundation which is shown in the elevation. (Courtesy of the Institution of Civil Engineers.)

The larger of the two was the Puentes dam; it was 50 metres (164 ft) high, 282 metres (925 ft) long, nearly 11 metres (36 ft) thick at the crest and 44·25 metres (145 ft) thick at the base (Fig. 16). Here then was a dam of gigantic proportions compared with anything that had been built before. Not only was it 30 feet higher than the Alicante dam, but it contained something like seven times the volume of material. It was not curved like the Alicante dam, but neither was it straight. In plan view it comprised three straight sections, 53 metres (174 ft), 124 metres (407 ft) and 105 metres (345 ft) long at the crest. So the Puentes dam can be regarded as a gravity dam of polygonal outline, its shape being dictated principally by the topography of the site.

The water face of the dam was vertical throughout its length. The upper 16·7 metres (55 ft) of the air face were made up of four steps, 4·175 metres (13¾ ft) high and 3·34 metres (11 ft) wide, and below this the face sloped uniformly down to the base. The structure was built with the usual core of rubble masonry set in mortar and faced with large cut stones. The crest was finished off with an ornamental parapet on which it was planned to place colossal statues of Charles III and his son Charles IV in whose reign the dam was completed. Even without these final ornamental flourishes, it must have been an imposing piece of work.

The dam was equipped with a large scouring gallery 22 feet wide by 25 feet high, so large in fact that at its upstream end the tunnel was divided by a central pier in order to allow the use of closing beams of manageable length. Water could be drawn from the reservoir through either one of two outlet wells. Both of these were at the centre of the dam, one on each side of the scouring gallery. They connected with horizontal galleries; one was at foundation level and the other about 100 feet below the crest. The outlet wells were much bigger than those used in earlier dams—14 feet wide by 8 feet—and they connected with the reservoir by means of holes. Each aperture was 1 foot wide by nearly 2 feet high and they were arranged in sets of three, side by side. The vertical distance between each set was 2¾ feet and there was, therefore, ample provision to draw water from the reservoir.

Apparently overflow was intended to discharge over the crest of the dam, something which, as things turned out, never happened. There is a suggestion, however, that a separate side spillway was built. On the left bank near the site of the dam are the remains of a channel, about 20 feet wide and 5 feet deep, which is not now used but would have been at roughly the level of the dam's crest. If not intended as a spillway, its purpose remains a mystery.

At first sight the place chosen for the Puentes dam is an excellent one, and so it must have appeared to the engineers. At a narrow point in the Guadalentin's valley, with hard rock on both sides, a dam 50 metres high would have created, when full, a reservoir 2½ miles long and containing perhaps as much as 3,260 million gallons (12,000 acre-feet) of water. Early on in the dam's construction, however, a deep pocket of earth and gravel was encountered in the foundations under the centre of the structure. At this point the project could either have been abandoned or moved to another place, but in fact the engineers, committed no doubt to getting the dam finished, elected to continue. They decided to carry the structure over the weak spot on a foundation of piles.

Hundreds of wooden piles, 20 feet long by about 2 feet square, were driven into the pocket of earth and gravel. They were connected and braced at their caps with a network of horizontal beams at right angles. Into this grillage the lower courses of masonry of the dam were built to a depth of 7 feet. The area of earth and gravel extended well downstream from the base of the air face, exactly where the scouring gallery and one of the outlet tunnels were intended to discharge. So, to prevent under-mining of the foundations below the dam, the piled grillage was extended 130 feet downstream. This apron was covered with 7 feet of masonry plus a layer of planks to protect the masonry from erosion.

Why the engineers chose the above way out of their problem rather than a complete excavation right through the soft material and down to bed-rock is difficult to say. The evidence is that Spanish dam-builders in general were fully aware of the need to found dams on rock. To what extent the Puentes engineers regarded the piled foundation as risky we shall never know; they may conceivably have believed their solution to be entirely feasible.

There is no doubt that the measures taken to overcome the foundation defects were extremely well executed. For eleven years—from 1791 until 1802—the dam was perfectly sound, but during this period the reservoir was never filled; the demand was so great that not more than 82 feet of water ever built up behind the dam. And then at the beginning of 1802 the water level began to rise and by April 30 had reached 154 feet. This was more than the foundations could take. The Puentes dam is the first serious dam disaster of modern times and one for which we have an eye-witness account. It makes fascinating if awesome reading.[33]

About half past two on the afternoon of 30 April 1802, it was noticed that on the downstream side of the dam, towards the apron, water of a very red colour was bubbling out and spreading in the shape of a palm-tree. Immediately someone was sent to inform Don Antonio Robles, the director of the works. About three o'clock there was an explosion in the discharge-wells that were built in the dam from top to bottom, and at the same time the water escaping at the downstream side in-creased in volume. In a short time a second explosion was heard, and, enveloped by an enormous mass of water, the piles, beams and other pieces of wood which formed the pile-work of the foundation and of the apron were forced upwards. Immediately afterwards a new explo-sion occurred, and the two big gates that closed the scouring-gallery, and also the intermediate pier, fell in; at the same instant a mountain of water escaped in the form of an arc; it looked frightful and had a red colour, caused either by the mud with which it was charged, or by the reflection of the sun. The volume of water which escaped was so considerable that the reservoir was emptied in the space of an hour. The water reached Lorca ahead of the messenger sent to tell the director of the first signs of the disaster; as the flood caught up with him, he was obliged to escape up a nearby hill.

The dam presents since its collapse the appearance of a bridge, whose abutments are the work still standing on the hillsides, and whose opening is about 17 metres broad by 33 metres high. At the moment of the accident the effective depth of the water was 33·4 metres. Its surface was 46·8 metres above the base of the dam; the lower 13·4 metres being taken up by the deposited material.

From this description the cause of the failure is clear. The dam in itself was sound enough, as its ample dimensions indicate. But once the pressure of 154 feet of water had blown a hole clean underneath the dam, the immediate collapse of a substantial part of it was inevitable.

The flood which hit Lorca, only twelve miles away, must have been colossal. The whole reservoir, perhaps 2,000 million gallons of water (the reservoir was not full), was discharged, we are told, in an hour, and there was little chance for it to distribute itself to left or right of the river valley. According to an official statement drawn up at the time, 608 people were drowned, among whom was Don Antonio Robles, the director of the works. The number of houses completely or partially destroyed was 809.

Eventually the remains of the Puentes dam were demolished and a new structure was erected. For eleven years the old dam had been the highest in the world. In 1884 this pre-eminence returned to Puentes, the second dam on the site being 71 metres (233 ft) high. Just above it some remnants of the first dam can still be seen—masses of rubble masonry and huge stone blocks, the left-hand abutment of an enterprise which was doomed from the outset.

The other eighteenth-century dam on the Guadalentin attributable to Charles III was nine miles upstream from the Puentes dam. Considering that it was a parallel development, it is interesting that the so-called Valdeinfierno dam (Hell's Valley dam) had a profile quite unlike that of its neighbour. It was 35·5 metres (116½ ft) high with a crest 12·5 metres (41 ft) thick. Four and a half metres below the crest the thickness increased sharply to nearly 30 metres (98 ft) and the dam's base thickness was 41·75 metres (137 ft). It was the most massive dam to have been built in Spain.

The structure was nearly 300 feet long at the crest and in plan view

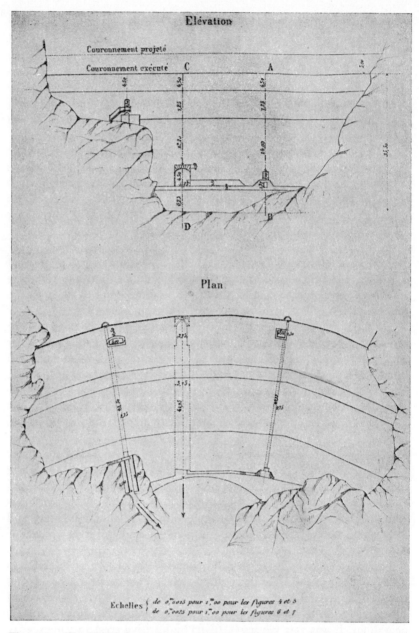

Figure 17 The Valdeinfierno dam as drawn by Maurice Aymard. As the elevation suggests the dam was intended to be higher and this was realised in 1897. (Courtesy of the Institution of Civil Engineers.)

comprised seven straight sections, a polygonal arrangement which amounted virtually to a circular form. The dam was constructed of the usual combination of materials and was founded entirely on rock. Structurally it was a complete success but hydraulically it was not.

There were two outlet wells near the water face, one feeding a low-level tunnel and the other, a tunnel half-way up the dam at the left-hand end. The vertical distance between the openings which connected the outlet wells to the reservoir was 3 metres (10 ft), and this was too great. When silt deposits closed the apertures at one level, no water could be drawn off until the water-surface had been raised ten feet. Consequently the front face of the outlet wells was demolished, and this must have worsened matters because silt must have begun to choke the outlet works. Perhaps this was a contributory factor in the dam's demise in the early part of the nineteenth century. Another was the opposition of local land-owners who complained that, whenever the dam was desilted, the sediment and water which poured down the river was damaging to their property. Nor must the fear of another disaster be discounted. As can well be imagined, the failure of the Puentes dam made a deep and lasting impression on the people of Lorca. Anxious not to repeat the experience, the local people perhaps abandoned the dam deliberately.

Anyway, by the middle of the nineteenth century it was completely disused and full to the brim with silt. It is interesting to note that in not much more than fifty years a large reservoir had been completely taken over by sediment, a clear indication of the magnitude of this problem. Near the end of the nineteenth century a new dam replaced the old one and is still in use.

Earlier it was noted that Charles III's reign saw the construction of numerous canals. Not all were for irrigation. A number of navigation canals were built or projected, and among them was one intended to connect the River Guadarrama at Torrelodones to the River Tagus at Aranjuez. Work actually started on the project in the last year of the king's reign, 1788, with the construction of a dam on the Guadarrama.[34] This was the most ambitious project of all and, had it ever been finished, it would have attained the stupendous height of 93 metres (305 ft).

The design of the dam was woefully inadequate and quite out of character with so much of the excellent work that had preceded it. The structure was intended to be 251 metres (825 ft) long, 4 metres (13 ft) thick at the crest and 72 metres (236 ft) at the base. The air-face and water-face walls were a mere 2·8 metres (9¼ ft) thick and connected by means of transverse masonry walls. Into the compartments thus formed the builders packed a mixture of rubble and clay. For eleven years this extraordinary dam slowly grew up and then met the fate which one might almost say it deserved. The position is neatly summed up in the following brief account:[35]

> On 14 May 1799, when the construction had already reached a height of 57 metres, there was heavy rain which soaked the clay and caused it to swell. A part of the front wall was overturned and the work has not been continued.

What is remarkable is that it actually reached a height of 57 metres (187 ft) before its career was brought to such an ignominious end. It must be the only dam ever to have failed by the direct action of rain rather than through the medium of a river.

We have seen, then, that around the end of the eighteenth century (1799 and 1802) two big dams in Spain failed and another went out of use. These facts, together with the ravages of the Peninsular War, not surprisingly cooled the Spaniards' enthusiasm for big dams and robbed the period of a good deal of its new-found technological momentum. However, at least one beneficial result can be detected. The School of Civil Engineering in Madrid was opened in November 1802, and the failure of the Puentes dam seven months before, together with its tragic consequences, was a stimulus to its foundation.[36] It was one of Europe's earliest institutions devoted to the teaching and development of the subject.

In the first half of the nineteenth century Spanish dam-building paused. The dams of Nijar and Lozoya, built around 1850, mark a revival, but now other European countries were also busily engaged in building dams, at first mainly for supplying canals but increasingly for public water supply as the century wore on. Never again was the limelight to be focused so permanently or so brightly on Spain.

This chapter can be regarded as the focal point of our story. Spain's record of dam-building is longer, more continuous and, before 1850, more important than any other country's. It was here that the Romans did some of their best work; the Moslems, although limited to river dams, at least laid the foundations of an irrigation technology which the Spaniards subsequently linked to the idea of large reservoirs. Christian Spanish engineers proceeded to build gravity dams, arched dams, arch dams, a buttress dam and, in a primitive form, the multiple-arch dam. The only post-Roman attempt at an earth dam was a failure. Irrigation dams predominated, but power dams made an appearance in the eighteenth century, and water supply dams in the nineteenth. Spain, in short, was the birthplace of modern dam-building.

Many nineteenth-century writers confirm Spain's position. They were not for the most part historians; rather they were civil engineers attempting to discover how the related arts of dam-building and irrigation could be advanced. That they looked to Spain for a model of what could and should be done surely indicates where the chief source of information and example was to be found. Many of the sources quoted in this and the preceding chapter typify the feeling that the lessons of Spanish irrigation would be of value in other countries. Maurice Aymard's book, for instance, was commended in its preface to the Algerian Governor-General; Sir C. R. Markham's interest in irrigation sprang from his concern for the agriculture of India, to whose government he submitted a lengthy report based on his findings in Spain; C. C. Scott-Moncrieff's visit to Spain was officially sponsored by the Indian Government; and Andres Llaurado knew full well that Spain's own future depended on progress along already established lines.

In later chapters we shall see the extent to which other European countries built dams on their own account and then the degree to which Spanish dams affected developments in the nineteenth century. That Spanish dam-building was 'exported' is clear enough, the best example being, as one might expect, in America.

6

The Early Americas

THE HISTORY OF dam-building in America begins in the pre-Columbian period. Easily the best-known and certainly the most important pre-Columbian civilisations in Central and South America were those of the Aztecs in Mexico, the Mayas in Guatemala and Yucatan and the Incas in Peru. However, this famous trio were by no means the only peoples to flourish in the days before Columbus, and only one of them, the Mayas, can be said to have monopolised the region it occupied for a long span of time. The Aztecs and Incas made their appearance at a very late date and each enjoyed only a short period of supremacy; and whereas the Maya civilisation went through its own natural cycle of rise and fall, the Aztecs and Incas were cut off in their prime, perhaps before it, by the arrival of the Spaniards.

The early history of the American continent is vague. The first inhabitants migrated into America from Asia in around 25000 B.C. and this movement continued until 7000 B.C. or thereabouts when the land bridge disappeared. From then on the 'Americans' were on their own and had no contact with other continents until the Europeans arrived.

About 1000 B.C. the first identifiable American tribes appear. They were builders in stone, practised agriculture of a sort and were capable weavers.

The period of the first 1,000 years A.D. is generally known as the 'classical period'. The Maya civilisation reached its peak during this millennium, the Incas and Aztecs had yet to appear, and everywhere agriculture was the basis of economic life. Metallurgy was being practised, and building in stone, often on a large scale, was common.

It is interesting that the technical history of pre-Columbian America[1] shows obvious parallels with the Old World and also certain very basic differences. Agriculture and irrigation, as we shall see, emerged along very similar lines, although the crops themselves were different. Textile-weaving shows a marked resemblance, pottery was manufactured to a high standard but without the use of the potter's wheel, water transport was an equally basic feature, and building in stone and mortar produced some fine structures but always without the use of the circular arch: only the corbelled arch is found in pre-Columbian America.

An even more remarkable absentee is the wheel. It was never used for vehicles, for pottery-making or in any form of rotary quern, yet it has been found in Mexico on toy animals of the period. Probably the absence of wheels is to some extent explained by the lack of a draught animal. The llama can only be used as a pack animal, and by the Mayas was not even used for that purpose.

Metallurgical developments were restricted. The Mayas were without metals, except gold, and even that was introduced only at a late date; the Aztecs reached the stage of using copper; the Incas had bronze but only on a small scale. So far as they went, pre-Columbian metallurgists were as skilled as their Old World counterparts. But they had no iron, and for this reason their technological development will not bear comparison with that of the Old World.

South American agriculture began in the mountain regions of southern Peru and northern Bolivia; the rainfall here is small but sufficient to allow such a development. Here, too, in a region which ultimately became the nucleus of the Inca empire, irrigation, while not essential, was at least desirable. When the people were forced to migrate to the coastal river valleys of Peru, irrigation techniques became absolutely essential; without them life could not have been supported. It is relevant at this point to note another basic difference between the Old and New Worlds. Old World civilisation grew up in the valleys of the big rivers—the Nile, Euphrates, Tigris, Indus and so on. In the New World it did not. It was not in the valleys of the Amazon or Plate that civilisation appeared. Instead it happened, in the case of South America, in the semi-arid highlands and arid coastal valleys traversed by small rivers; and in Central America the Mayas, the Aztecs and the latters' predecessors were not river civilisations at all.

The northern coastal valleys of Peru were populated first by the Mochicas (A.D. 400–1000) and then by the Chimu (1000–1466). These were the peoples who developed the irrigation of the area, especially on the rivers Chicama and Moche near the ancient city of Chan-Chan. Many hundreds of miles of irrigation canals, made for the most part of adobe, have been located; they once watered areas which today have reverted to desert. Some of the main canals were large and long. One has a total length of 87 miles and a cross-section 8 feet wide by 6 feet deep;

another, 70 miles long, has an evenly graded fall of 4,000 feet corresponding to a gradient of about 1 in 90.

The systems of main canals fed many dozens of branch canals. For the most part the entire canal systems followed contour lines, so that bridges and cuttings were rare. However, near Ascope a huge adobe aqueduct bridge, half a mile long and 40 feet high, is still visible.[2] The canals were, of course, the only way of directing water to the fields because no water-raising devices were known. In some cases these ancient canals are still in use.

The canals were fed from dams across the rivers.[3] In some cases the same river was dammed at different places in order to provide water at different elevations. The dams were made of earth and equipped with side weirs to divert water into the canals. It is not clear what steps were taken to allow a dam to pass water downstream to the next structure.

At this stage an important point must be made. The irrigation schemes of coastal Peru are technologically similar to those of North Africa and the Middle East. They emphasise how basic a technology irrigation is, and indicate that very often, even though place and period are different and contact entirely absent, man will arrive at the same solution to a given technical problem when it is posed in similar conditions. Furthermore, we find once again that when large irrigation schemes are needed the construction of the required dams and canals can only be undertaken by large social groups who are well organised.

In addition to diversion dams, Peru's early inhabitants also constructed reservoir dams. A few of these were in the foothills of the Andes, the remainder in the coastal areas. These appear to have been designed to impound the surplus water which flowed when the snows of the lower Andes melted in the spring. One of these reservoir dams was built in the Nepeña valley and impounded a lake three-quarters of a mile long by more than half a mile wide.[4] In this case the dam did not span a river. Where the Mochica engineers found a natural depression in a dry valley, they enlarged the space by excavation and built a dam of rocks and earth across one end between two rock faces. The dam's dimensions are not given beyond the fact that its base thickness was 80 feet. Water was directed to the reservoir by means of two canals. One of these was fourteen miles long and carried a supply from a river; the other was five miles long and fed by springs.

In 1466 the Incas overran the Chimu empire, this being the final stage of their expansion. Fine engineers though they undoubtedly were, especially in bridge- and road-building, no dams can be attributed to them. Perhaps their reign in territories where dams were feasible and desirable was too short. By 1535 the power of the Incas had been destroyed by the Spaniards, and they, despite their experience in their home country, did nothing to preserve or extend any of South America's irrigation works.

The Mayas had a water problem of some magnitude. The Yucatan peninsula and the hill zone to the south are not lacking in rainfall but are decidedly short of the means to use it. There are very few rivers (none at all in the north), and the very thin layer of topsoil is incapable of holding water for long. It quickly soaks away into the limestone substrata, itself very porous.

Although the Mayas did not practise irrigation, they did have to tackle the question of water supply to their several large cities. They worked out three solutions: the artificial well (*cenote*), the underground cistern (*chultun*) and the large open reservoir (*aguada*). *Aguadas* were built by the hundred and were sometimes very large. John L. Stephens has left details[5] of a few which he found in the nineteenth century. They were constructed in natural depressions which had to be carefully lined with masonry to prevent immediate loss of the water into the soft and porous limestone. It was not uncommon practice, apparently, to build *cenote*s and *chultun*s into the depths of the *aguada*s. These could then be utilised if the *aguada* dried up.

The extensive networks of Mayan *aguada*s represent civil engineering on a grand and well-organised scale; but the extent to which their construction involved dam-building is not at the moment very clear. The best-known example of a Mayan dam is at Tikal, the biggest of their cities and one frequently affected by droughts despite being in a very wet region. An entire ravine was fitted with a waterproof lining and then dammed to form a reservoir.[6] The masonry dam's crest also acted as a bridge.

The Mayas' failure to develop irrigation may well have accelerated their decline. Whether or not dams would have helped is doubtful, but rotary motion probably would have done. There was a lot of water to be had from the limestone aquifers, and the use of wells and cisterns suggests that the Mayas knew this. What they did not know was how to raise it to the surface by even the most elementary of mechanical devices.

The Aztecs' claim to hydraulic fame is well known already. In Mexico they were the only people to develop irrigation or grapple with hydraulic problems. In 1325 the Aztecs settled on an island at the western edge of Lake Texcoco, one of five lakes which stood in the Valley of Mexico.

Agriculture in and around their capital city, Tenochtitlan, was based on the ingenious *chinampas* system,[7] a technique which could only thrive on supplies of fresh water. Unfortunately frequent floods in Lake Texcoco swamped the irrigation system with salt water as well as disrupting life in the capital by laying it under water. In 1449 the Aztecs took decisive action.

To the east of Tenochtitlan, at the narrowest point of Lake Texcoco, they built a dyke, 10 miles long and 20 feet high. It consisted of a masonry-and-earth core set between outer walls of wooden piles joined together with sticks. This enormously long structure was never required to resist any great pressure since the water levels on each side probably never differed by more than a few feet. Nevertheless it was a considerable undertaking and required a labour force, so it is said, of 20,000 men. Even so, the future safety of Tenochtitlan and its irrigation system was not ensured for ever, as the Spaniards were soon to discover.

The Spanish conquest of the Inca empire was effectively complete by 1535. Probably the one thing that impressed the conquistadors of Peru more than anything else was the enormous quantity of gold and silver that they found there. Every ounce that they could lay their hands on was shipped back to Spain in compliance with Charles V's edict of 13 February 1535: 'All gold and silver from Peru shall be melted in the royal mints at Seville, Toledo and Segovia.'

Ten years later the Spaniards were introduced to the 16,000-foot-high 'silver mountain' of Potosí, 300 miles south-east of La Paz. The *cerro rico* proved to be one of the biggest and richest deposits of silver in the world and provided Spain with the bulk of her total supply for nearly a century. So great was the output that silver overtook gold as the main component of the 'treasure' which crossed the Atlantic.

The city of Potosí grew up near the mountain at an altitude of nearly 14,000 feet. It was then, and has been ever since, one of the highest cities of any size in the world. At the height of its prosperity (around A.D. 1600) it may have had a population of as many as 160,000 people; today it has only 45,000. The climate in the area is cold and very dry, something which the Spanish conquerors were quick to notice. The annual rainfall amounts to about 25 inches, and the bulk of this falls in the three summer months, January to March. Furthermore, such water as falls is subject to a tremendous rate of loss by evaporation because the atmosphere is so dry and rarefied.

For twenty years producing high-quality silver at Potosí was a simple task. Hundreds of galleries driven into the mountain yielded the richest of ores which needed only primitive smelting furnaces to convert them into ingots of pure silver. By 1566 the veins of rich ore were exhausted, but the demand for silver continued and may in fact have increased. In order to work the poorer ores it was essential to crush them first, and for this purpose it became clear that water-powered grinding-mills were a necessity.

The problem was discussed at a meeting held in Potosí in 1572. The nearest permanent river capable of powering the proposed grinding-mills was ten miles away, and it was decided therefore to construct a reservoir near the mines, which would impound the summer rains for use during the remainder of the year. Four local miners agreed to finance the project, the Spanish viceroy offered to provide a labour force of 20,000 Indians, and this offer was finalised by Philip II in 1574. Work on the Chalviri dam[8] began immediately.

Soon there were problems. The grinding-mills were built near the dam, and this resulted in the ore having to be moved about far too much on the backs of some 2,000 llamas. A better technique was to build mills at the mines and in the city itself, and then channel water from the dam to the mills. No sooner had Potosí been furnished with an aqueduct and thereby a vastly improved water supply than its period of greatest prosperity and activity began. A total of 132 ore-crushing mills were eventually built in the city and all were fed from a single water channel whose total fall was 594 metres (1,950 ft).

William E. Rudolph has estimated that an average flow of nearly 9 cubic feet per second was available to drive the mills. He has estimated, too, that the total power output was 600 h.p., but one would have to know a good deal about the size and form of the water-wheels to check these figures. One thing is certain, however: Potosí was a silver-mining centre of some size and considerable output and an early example of a city which thrived on water power.

By itself the first reservoir, called Chalviri, was inadequate to meet the needs of 132 mills. As the city's industry expanded more dams had to be built, and by the first quarter of the seventeenth century there were thirty-two of them, the total capacity of the reservoirs in their final form being about 1,320 million gallons (4,850 acre-feet). Such an elaborate system of reservoir dams obviously demands our attention, and fortuately a large number of them survived into this century. When the system was rebuilt in the 1930s much about the dams' construction was elucidated.[9]

Potosí is flanked on its eastern side by a mountain range whose axis runs north to south. Seven glacial troughs radiate towards the city, and along these the dams were built in series. Basically there are two different types of reservoir.

At the heads of the glacial valleys, the Spanish engineers sought out lakes which were naturally dammed by rock shelves. Along these shelves they proceeded to construct dams in order to increase the lakes' capacities. The dams are equipped with low-level outlet tunnels through which to draw the stored water. These reservoirs have always contained a volume of 'dead water' below the level of the outlets. Apparently no attempt has ever been made to pierce the natural rock dams.

Further down the valleys the reservoirs are wholly artificial and depend entirely on dams for their capacity to impound water. At these sites the beds of the reservoirs and the foundations of the dams are soft moraines. As a result, the lower reservoirs have been prone to leakage and a good deal of the water collected is soon lost. Furthermore, the lower reservoirs have suffered badly over the years from siltation. Water flowing over the loose moraines has eroded large amounts of sediment and carried it to the reservoirs. One of them is completely full of silt, and several others have lost a considerable proportion of their storage capacity.

By contrast the rock-bound upper reservoirs are a better proposition. They are not prone to excessive leakage, and evaporation, although excessive compared with what would occur at sea-level, is relatively less because the reservoirs are more sheltered from the wind and the air temperature is lower.

The natural advantages of the upper reservoirs were evident to the Spanish engineers. It is significant that it was these sites which were used first, while those in the lower reaches of the valleys were only employed as a last resort.

The viceroy who first projected the idea of dams at Potosí was one Francisco de Toledo. Although he himself only gave approval for eighteen reservoirs, fourteen more were built. Of the complete set, a total of twenty-four were still in service in 1936. One other, the reservoir of San Lazaro, was the one full of silt and out of use.

The manner of using the reservoirs is perfectly clear. In the wet season all were allowed to fill right up. On any given valley, the upper reservoirs were then used to feed the lower ones as water for the mills in Potosí was drawn down during the year. From the lowest dam on each valley an aqueduct was built, and these were all brought together near Potosí. Their combined flow was then channelled into the main canal which supplied the city's mills.

It would be tedious and repetitious to detail each one of the dams of Potosí. Instead it will be enough to describe the two basic types of dam which were used, with examples.

The biggest structures—those impounding the greatest quantities of water—were the most substantial. Each of these dams had a cross-section made up of five distinct structural elements. The water-face wall was about 75 centimetres ($2\frac{1}{2}$ ft) thick and made of masonry; the joints were not mortared. This portion was clearly only a protection against wave action and was not a watertight skin. Next there was a section of clay fill which presumably was intended to be watertight and in most of the dams was a few metres thick. The core of each dam was a masonry wall about 1 metre ($3\frac{1}{4}$ ft) thick at the crest and increasing to as much as 2 metres ($6\frac{1}{2}$ ft) thickness at the base in the bigger dams. This part of each dam was cemented with lime mortar and must be regarded as the main water-retaining element. Behind the central core was another layer of clay fill a few metres thick, and this was held in place by an air-face wall of masonry blocks and lime mortar. The air-face wall and the clay fill it supported were not as high as the other three layers.

The dams which can be termed 'large' were built to a height of 6–8 metres (20–26 ft) and were 10–12 metres (33–40 ft) thick. Their lengths ranged from 64 metres (210 ft) up to 500 metres (1,640 ft). The reservoirs had capacities which varied from about 9 million gallons (33 acre-feet) up to some 95 million gallons (350 acre-feet); their perimeters were anything between one and two miles in length.

One of the reservoirs was much bigger than all the others. The Chalviri reservoir, whose dam (Pl. 33) was 8·5 metres (28 ft) high, 10–12 metres (33–40 ft) thick and 226 metres (740 ft) long, had a storage capacity of nearly 600 million gallons (2,200 acre-feet). In other words this one reservoir comprised nearly half the total capacity of the whole system. It is of some interest to note that the first reservoir built was by far the biggest, and this emphasises how all the others were added piecemeal as the demand for water gradually increased.

Large dams of the type described above accounted for about one-third of the total. The remainder were small dams, 3 metres (10 ft) or less in height, and of much looser construction. They consisted essentially of twin masonry walls, sometimes with mortar but often without, and between them a filling of clay and earth. Over the centuries it is these structures which, not surprisingly, have been most prone to deterioration and leakage.

When many of the dams were being rebuilt thirty years or so ago, other features of their construction were elucidated. The foundations, for instance, were invariably found to be badly made; the cut-off trench was less than 1 metre (3¼ ft) deep and throughout their life the dams were plagued by excessive leakage as a consequence. The outlet valves in the dams were simple and inadequate. On the crests of many of the dams there was a horizontal beam which could be used to raise and lower a pole inside a vertical well. At its lower end the pole was equipped with a sliding plug which fitted into the outlet pipe. On the later dams a more sophisticated technique was used. It consisted of an inverted bronze cone on a chain, which could be raised and lowered within the mouth of a vertical standpipe on the dam's water face. Unfortunately both these types of 'valves' were very difficult to use once stones and debris had been trapped within them, and this happened all too easily. Finally, there was no evidence that any of the dams had spillways. Because of their mode of construction it was a certainty that floods would cause damage, and over the centuries this happened often.

Whether it was the lack of a spillway or whether it was caused by water undermining its foundations, the San Ildefonso dam failed completely on 15 March 1626.[10] It was one of the biggest dams, 8 metres (26 ft) high and 500 metres (1,640 ft) long, and formed a reservoir of 95 million gallons (350 acre-feet), second in volume only to that of the Chalviri reservoir. In the middle of March it must have been full or very nearly so.

Within two hours of the collapse, the entire contents had ripped through Potosí only two miles away. It is said that 4,000 people were killed, but this may be an over-estimate. On the other hand the destruction, total or partial, of 126 of the 132 mills is a figure on which contemporary records can probably be trusted; it must have been one of the most serious aspects of the disaster and one of the easiest to quantify.

Just as Potosí had grown famous and prosperous from its dams, so the failure of one of them inaugurated its decline. More than half the mills (79) were never rebuilt, and even though the San Ildefonso dam and reservoir were put back into service, Potosí never really recovered. Miners the world over are very superstitious, and the accident at Potosí was taken to be an act of divine retribution which had to be heeded. Other more rational factors played a part as well. Decades of intensive mining had depleted the veins which were easy to work, and finding new veins gradually became more difficult. Moreover the discovery of the mercury amalgam process enabled other sources of silver to be used economically for the first time. (It also initiated more intensive working of the notorious and horrible mercury mines at Almadén.)

The water-works of Potosí were a fine piece of engineering planning, although the dams themselves left something to be desired both structurally and hydraulically. Even so, and despite centuries of neglect, it was possible to resurrect them in modern times as part of Potosí's new attempt to win prosperity from metal-working. The new 'precious' metal is tin. The dams of Potosí can also claim a curious distinction: built at altitudes of between 14,000 and 15,000 feet, they are probably the highest in the world.

The Spanish conquest of Mexico was a sad and bloody business, and fortunately we do not have to discuss it. Some attention, however, must be given to the Spaniards' inheritance of the Aztec capital and its various hydraulic problems, to which they were introduced at an early date.

In 1553, 1580, 1604 and 1607 there was severe flooding of Mexico City (formerly Tenochtitlan). Following the flood of 1607, a Dutch engineer, known in Spanish as Enrico Martinez, attempted to drain the northern lakes by means of a tunnel four and a half miles long.[11] He failed. Another Dutch engineer called Boot built further dykes and embankments which were partially successful. In the 1620s Martinez tried once more to build the tunnel, and once more he failed. This time he was sentenced to five years in the dungeons. In 1634 he was reprieved and instructed to tackle the flood problem again. His answer was the construction of a dyke to separate Lake Texcoco from the two northern lakes which were those that collected the bulk of the seasonal rain.

The so-called Calzada of San Cristobal was a solid piece of work, four miles long, 28 feet thick and 10 feet high.[12] It was made of masonry and buttressed throughout its length on both sides. At three points there were sluices to regulate the flow of water in both directions. However, like its predecessor, it was not required to resist great pressures and can only be regarded as a dam in a marginal sense.

The Conquistadors had come from a land of dams and irrigation canals. In Mexico they found themselves in a country that was ripe for the introduction of their expertise. As soon as the brief period of fighting and bloodshed had passed, the religious factor quickly asserted itself. But the Catholic fathers did more than preach (although they did plenty of that); they also played the parts of teacher, doctor, lawyer, architect and engineer. Essentially it was the influence of the Church which enabled Mexico to acquire Spanish irrigation technology.

An early enterprise was due to one Padre Juan de San Miguel, a Franciscan. To the north-east of Mexico City, in the plains of Uruapan, he dammed various small rivers so as to divert water into canals.[13] In recent times (the 1930s) this irrigation system was brought back into use to meet local agricultural requirements.

In southern Mexico a contemporary of Padre Juan, Padre Francisco Marin, carried out similar irrigation works in order to divert water from relatively wet areas to those that were arid.

In 1548 it was an Augustinian, Padre Diego de Chavez y Alvarado, who began the formation of the Lake of Yuriria.[14] This huge reservoir, ten miles long and four miles wide, was a natural formation needing no dams. But to fill it with water required the construction of dams across a tributary of the River Lerma from which the required supply could then be channelled. The reservoir was also coupled to another adjacent lake as an additional source of water. With alterations the Lake of Yuriria continues to be used.

In 1603 the Spanish Viceroy, the Marques de Montes Claros, author-ised the founding of the town of Salamanca in Central Mexico. It became the first new town in America whose establishment and growth was based on irrigation agriculture. In this case the canal system was fed by dams on the Lerma river and its tributaries.

Attention was paid to water-supply systems as well as irrigation works. To what extent this involved the construction of dams to divert river water is not known, but some fine aqueduct bridges were certainly constructed. There are three on the forty-mile aqueduct built under the direction of the Franciscan, Padre Francisco de Tembleque, between 1570 and 1587. The biggest of the three bridges has sixty-seven arches whose maximum height is 120 feet; it is a sure index of the state of civil engineering at the time.

Although the Spaniards utilised diversion dams and canal systems at an early date in Mexico, the large dam for water-storage was slow to appear. The earliest examples, so far as is known, date from the second half of the eighteenth century. They are mostly to be found in Central Mexico and especially in the state of Aguascalientes, 280 miles north-west of Mexico City. In this area there are apparently some twenty old dams, of which three are particularly interesting.[15]

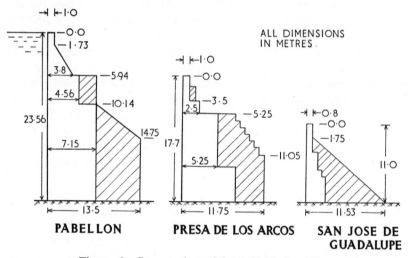

Figure 18   Cross-sections of three old Mexican dams.

The Pabellon dam (Pl. 34) stands across the river of the same name about twenty-five miles north of the city of Aguascalientes. Originally the dam was built to a height of about 58 feet and at that stage was nearly 500 feet long at the crest. The thickness at the crest was close to 15 feet. Some 14 feet below the original crest the dam's thickness increases in a single step to $23\frac{1}{2}$ feet, and below this sill the air face is vertical.

At some time the dam was heightened by the addition of a wall of triangular cross-section topped by a small parapet $3\frac{1}{4}$ feet thick. This extension brought the dam's total height up to 77 feet. Throughout its length the dam's water face is vertical, and thus the base thickness of the dam is $23\frac{1}{2}$ feet. Such a profile is inadequate for a gravity dam, and this accounts for the presence of a row of buttresses along the air face.

The buttresses are a somewhat random collection, non-uniform in size and unequally spaced. Each one projects about 21 feet from the face of the dam and is some 7 feet thick. At their tails they are 29 feet high, and their tops slope upwards to meet the body of the dam at a height of 44 feet. According to the figures given by Julian Hinds[16] there must be a total of some thirty buttresses. When the dam was raised it was the intention to raise the buttresses also. The rough and uneven finish on the tops of most of these indicates that this work was never completed. Apparently, too, the extension wall is also incomplete: the left-hand half is lower and of rougher construction than the section on the right.

The dam is fitted with a masonry spillway at the far end, but it is never used. In theory the overflow level can be adjusted with stop-logs, but in practice all are kept permanently in place. Consequently, overflow normally pours over the dam's crest at the left-hand end. In times of exceptional flood the whole of the left half of the dam's crest, which is a few feet lower than the right-hand half, becomes a spillway.[17]

The average annual run-off of the Pabellon river is about 40,000 acre-feet, and the flow at the dam site is very flashy, being negligible for a large part of the time. The reservoir's capacity is about 800 acre-feet, so that it is easy to keep it full most of the year. On the other hand the reservoir is too small to act as a protection against floods. The dam was in fact built to meet the twin needs of irrigation and power.

For irrigation requirements the dam has two outlets. One is a low-level sluice which used to feed a canal leading off to land below the dam. Today the outlet is permanently closed with wooden plugs and there is no trace of the apparatus which must once have controlled the sluice. The second outlet diverts water to a pair of supplementary storage-tanks just below the dam. Presumably these were used to sustain irrigation if the reservoir ever ran dry.

Half-way down the dam's air face at the right-hand end there used to be a flour-mill. Its remains can still be seen although most of its equipment, including the water-wheel, has gone. It is not clear how the flow of water from the reservoir to the water-wheel was controlled. The tailrace from the mill joins the low-level outlet canal just below the dam. The Pabellon dam appears to be among the first in Central America designed to store water for a hydro-power plant.

The reservoir is appreciably silted, and there is no means to flush these deposits away in the manner of the old Spanish dams. Consequently the low-level outlet probably went out of use at a very early date, leaving the high-level outlets to the mill and the tanks as the only usable ones. Undoubtedly the accumulation of sediment was the reason for the dam's being heightened and may also have initiated the construction of the supplementary tanks.

The dam known as the Presa de los Arcos (the Dam of the Arches) is on the Rio Morcinique, seven miles east of Aguascalientes. Its name has nothing to do with its shape (see Fig. 18 and Pl. 35). It is in fact another buttress dam, not so high as the Pabellon dam but longer.

In its present form[18] the Presa de los Arcos is 65 feet high and 720 feet long. The main wall has a vertical water face and a stepped air face. It is very likely that this dam has also been heightened at some time and that this work was never properly finished. All along its crest, 'starters' made of long pieces of rock stand ready to be keyed into more courses of masonry which were never built. Only at the right-hand end does the crest appear to have been completed.

Even the original buttressing is incomplete, as Plate 35 shows. The completed buttresses project 13 feet from the base of the dam which is itself 25½ feet thick. The ends of the buttresses have vertical faces 22 feet high and then are stepped upwards to within 8 feet of the air face of the main wall, which at this level is just over 17 feet thick. Above this level is the heightening wall, which has small buttresses supported on the flat tops of the original ones. The thickness of each main buttress is 8¼ feet, and this is also the width of each inter-buttress space. In other words half the dam's air face is buttressed.

The dam is in two straight sections of roughly the same length. Presumably it was so constructed to utilise the best foundation line, but the result is a dam whose apex points downstream. That the water-pressure has not split the dam open at the angle between the two sections says much for the tensile strength of the material at that point.

The Presa de los Arcos was an irrigation dam only; there is no evidence of a mill-house at the dam or anywhere near by. The dam has two low-level outlets. Both feed irrigation canals which are led away from the dam on small bridges. One of these canals, which is still in use, is carried across the river below the dam on an arched bridge; hence the dam's name. Nothing is known of the equipment which was originally installed to operate the outlet sluices.

The dam has an overflow spillway at the left-hand end. In the customary way its stop-logs are never removed and water pours over the crest and down the air face.

In the dam of San Jose de Guadalupe,[19] also on the Rio Morcinique, the buttressing technique is more fully developed (Fig. 18). Originally this dam was about 30 feet high with a vertical water face. Below the crest, 3 feet thick, the air face of the main wall is stepped so that it attains a maximum thickness of 9 feet at the base. This represents a much more slender dam than either of the others, and long buttresses were absolutely essential.

Each buttress is triangular in elevation, 28¾ feet long at the base and 30 feet high—in other words, reaching up to the crest of the original dam. Each one is 5 feet thick and they are spaced at 29¾-foot centres. Compared with its predecessors, the dam of San Jose de Guadalupe is a significant advance and a bold piece of construction into the bargain.

The dam is equipped with an overflow spillway at one end, but this is never used as such. The stop-logs, however, are not a permanent fixture in this dam. In the dry season they are removed to let water flow down the river-bed to supply irrigation channels.

An inscription on the dam states that it was heightened in 1865. This was achieved very simply with a parapet wall $5\frac{3}{4}$ feet high and 3 feet thick. It must have been built to increase the reservoir's capacity or to counter the effects of silt.

In his discussion of the above dams, Julian Hinds presents the results of a simple stress analysis of the structures—something which the original builders, of course, could not do. Only the dam of San Jose is theoretically stable against overturning, and on the basis of Hinds' stress analysis is the only one not subject to vertical tensile stresses at the water face, although there must be some horizontal tension in the main wall. The Presa de los Arcos is very nearly free of vertical tension, but the Pabellon dam sustains stresses of 70 lb per square inch. That the dams have successfully withstood a certain amount of tension and also the erosive effects of 200 years of overflow is testimony to the materials of which they are made.

All three consist of rubble masonry set in hydraulic mortar. None of the stone blocks have a cut finish, although many are roughly 'squared'. The dams have a generally smooth finish, but there are no specially fitted facing stones.

The mortar used in the dams' construction is of very high quality and contains locally manufactured hydraulic lime. Right back in the Aztec period a traditional building material was *tezontli*, a type of pumice, pink in colour and composed of silica and volcanic ash. Such a material would certainly behave like pozzolana and render lime hydraulic if added to the mixture. *Tezontli* or something very like it must have been used in these early Mexican dams in order to produce hydraulic mortar of such strength and durability. Indeed, the mortar is reported to be harder and stronger than the masonry which it binds together.

It is not known for certain how the dams are keyed to their foundations, although this was evidently done very securely. Julian Hinds' suggestion is that long pieces of rock were cemented into holes in the bed-rock and the dam was bonded to these 'dowels'.

None of the three dams described are fitted with scouring galleries. Nevertheless this technique was certainly known in Mexico at the time the dams were built. An example is the San Raphael dam in the state of Hidalgo, north of Mexico City (Pl. 36). The structure has not one de-silting tunnel but two, both very large. They appear to be about 20 feet high and 10 feet wide in a dam whose total height is perhaps 40 feet. Below them is a masonry apron to prevent erosion of the foundations when the sluices are opened up.

There is no lack of evidence that the Spaniards brought dam-building and its associated technologies from Spain to the Americas. Storage and diversion dams were built to feed irrigation schemes, to provide supplies of drinking-water and to power water-wheels, the latter being something entirely fresh in the New World. That the idea of buttress dams was current in Spain in the first half of the eighteenth century, and that the same type of structure should be found in Mexico in the second half, is especially indicative of the transmission that was in progress. In Spain the dam of Almendralejo is well made but of conservative design. In Mexico those of Pabellon, San Jose and the Presa de los Arcos are of inferior workmanship but structurally more daring. Like the Spanish dam, the Pabellon dam impounded water for use in a mill, and the materials of construction are common to all the dams.

It is interesting that arch dams do not seem to have been built in the Americas in the early centuries of Spanish rule. Scouring galleries, however, such a conspicuous feature of Spain's curved dams, are found in at least one Mexican dam; there may be others. It is a pity that nothing is known of the type of water-wheels used in conjunction with early Spanish–American dams, or of the form of such things as the sluice-gates in the big dams. Details such as these are often valuable in substantiating and illuminating cases of technological diffusion.

From Central Mexico, dam-building and irrigation was taken northwards to regions which are now part of the United States. In states such as New Mexico, Arizona and Colorado the idea of irrigation agriculture was not entirely new, however. Simple Indian schemes have been located in a number of places, especially in Arizona.[20] The techniques used are basically those which were characteristic of primitive irrigation in the Old World. Dating from the eleventh century there are examples in Colorado and Arizona of tiny dams built across intermittent streams for the purpose of accumulating soil and conserving water. Essentially these are the North American version of the wadi dams of the Negev. Not that there is any connection; it is simply that the technique is fundamental and inevitable under certain conditions.

In southern Arizona the Indians practised irrigation by canal networks, elaborate systems using main and secondary canals (just as in the Old World), many tens of miles in length. The oldest canal of any size goes back to about A.D. 800. The extent to which dams were used to feed these schemes with water is not certain.

Spanish penetration into the United States area began very early; it was inspired by the lure of precious metals and a zeal for religious conversion; and the latter was the medium through which dams came to be built. Following Father Eusebio Kino's occupation of lower California and southern Arizona around 1700, further expansion during the eighteenth century led to the establishment of missions along the Pacific coast and across the states of Arizona and New Mexico to Texas.

The water supply and irrigation needs of these missions led to the construction of dams. The San Antonio region of Texas was first settled by Jesuit missionaries in 1718, and in the same year irrigation canals were dug. In 1720 the San Jose Mission constructed an irrigation scheme which included a water-powered grist mill, while another canal, built in 1729, was large enough to be navigated.[21]

There is a suggestion that in laying out these canals levels were established by means of an A-frame device, from the apex of which a plumb-line was hung to coincide with a mark on the horizontal bar. In Moslem Spain such devices were well known, and that they were subsequently used in America is a plausible theory.

The diversion dams built around San Antonio were generally temporary affairs of loose rock and brushwood, but some solid ones of stone and mortar were also built, and that shown in Plate 37 is among the oldest dams in the United States still in use. Called the Espada dam, it was built across the San Antonio river in the middle of the eighteenth century. It is founded on a natural ledge of rock in the river-bed and stands to a height of about 15 feet.

In California, Jesuit fathers established missions along the coast in the last third of the eighteenth century. One of these took its irrigation and drinking-water from the Old Mission dam built across the San Diego river in 1770.[22] It was a long structure of rubble masonry set in mortar and was about 5 feet high. By the twentieth century the dam had been allowed to fall into ruins (Pl. 38). Near the centre a sluice with wooden gates was fitted, through which water could be drawn. The dam has now apparently disappeared.

A few years after the construction of the Old Mission dam, Jesuit fathers established an irrigation reservoir about ten miles east of Los Angeles. The El Molino dam was straight in plan, about 200 feet long and 15 feet high. It served to raise the level of a natural lake and to control the flow of water to the lands around the San Gabriel Mission. At the beginning of this century the dam was still in excellent condition but was no longer in use.

While Spanish influence was at work in the south-western corner of the United States, the north-eastern area was experiencing the development of dam-building of a different type and from a different source.

Northern European settlers in New England brought with them their own ideas about dam-building to an area whose requirements and resources were unlike those of the south-west. New England has little or no irrigation problem, because there is adequate rainfall which is well distributed throughout the year. The early settlers saw immediately that

F

there was sufficient water available to furnish them with a source of power.

In England, France and Sweden, the countries from which so many of the colonists came, small water-power dams were a familiar aspect of life at an early date. Depending on what materials were most readily available, they were built of masonry, brick, earth, timber or rock, or some combination of these, and supplied water at a modest velocity to simple water-wheels; either the horizontal wheel, sometimes known as the Greek or Norse wheel, or the vertical wheel of the undershot type.

It was the timber dam which quickly came to predominate in New England;[23] the readily available supplies of wood guaranteed this. One of the earliest dams was on the Piscataqua river at South Windham in Maine. It was built in 1623 to power a sawmill, the first, so it is said, in the United States. In 1634 another dam for the same purpose was built between Milton and Dorchester on the Neponset river in Massachusetts. In this case a grist-mill and a powder-mill were both part of the same installation. The engineer was one Israel Stoughton. At Portsmouth in New Hampshire another grist-mill was powered from a dam at an early date.

New England's early dams, small though they undoubtedly were, played an important part in the life of the communities who built them. The dams powered the sawmills which produced timber for building and the grist-mills which ground the grain for food. Naturally enough, towns grew up around the spots where these industries flourished.

Apparently nothing of these early dams has survived; they were, after all, made of wood. Their constructors were generally millwrights, as they had been in Europe for dams of the same type, and such men were not disposed to work from drawings or to leave written records or accounts of their work. However, their traditions lived on well into the nineteenth century, and the structures shown in Figure 19 are probably a good guide to the earlier dams.[24]

They were unsophisticated pieces of work formed basically from a collection of logs or beams laid at right angles to each other so as to build up a stout wooden framework. All manner of configurations were popular, and a common feature was a filling, between the wooden bars, of earth, stones, gravel or brushwood, or some combination of these materials. The fill was compacted as much as possible to give the whole structure weight, stability and watertightness.

On soft foundations of sand or gravel, the typical wooden dam was held in position by its sheer weight or, better still, by means of piles. Downstream aprons were provided to prevent overflow undermining the structures. On rock foundations, the lowest courses of timbers were anchored with iron spikes.

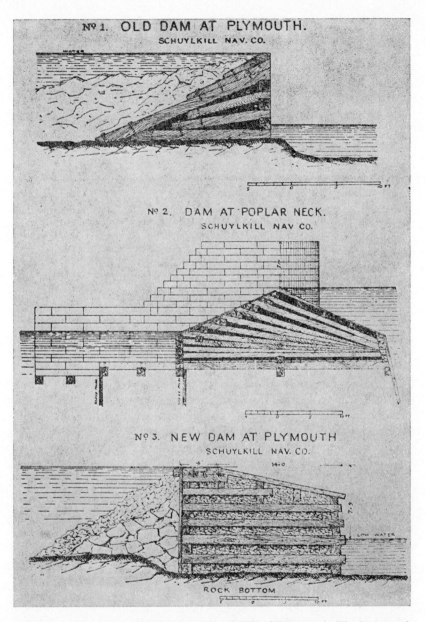

Figure 19 This picture is taken from Plate XCVI of Wegmann's *The Design and Construction of Dams*. It shows the cross-sections of three old American timber dams with earth and stone fill. (Courtesy of John Wiley and Sons.)

In most cases the water faces were long and sloping and covered with a layer of earth and gravel. The downward pressure of the water was thus inclined to stabilise the dam and hold the facing materials in place. It seems that the need to pass ice, logs and debris over the dams' crests with the minimum degree of obstruction was also a determining factor in the use of gently sloped water faces. The air faces were frequently vertical (see Fig. 19) so that overflow fell straight on to the apron or into the tail-water.

Prior to 1800 timber dams were perhaps up to 100 feet in length, usually about 10 feet high and rarely higher than 15 feet. At a later date, however, some much bigger ones were built. Generally timber dams were straight, although arched versions were not unknown in certain circumstances.

An interesting early use of dams in New England was in connection with certain tidal power schemes. The earliest example appears to be the one built on Mill Creek near Boston in 1631; and subsequently other tidal inlets along New England's coastline were also harnessed. There were, of course, earlier examples of tidal power plants in Europe, and it can safely be assumed that wooden dams like those built in Europe served to impound the incoming tide in the first North American versions. They were probably never very large.

The earliest masonry dam built in the north-eastern United States was due to English colonists. In 1681 they settled on the site of New Brunswick, New Jersey, and in 1743 they built a masonry dam across the Raritan river in order to provide a water supply. There may have been earlier masonry dams due to the Spaniards in the South-west, but otherwise the New Brunswick dam was the first of masonry in the United States. The dam had a long life: having been strengthened in 1780, it continued in use until it failed in 1888.[25]

Undoubtedly a lot of dams were built in North America before the nineteenth century. Important though they were to the processes of settlement and colonisation, they do not occupy an important place in the evolution of dam-building as such. Not until the nineteenth century did the need for big dams emerge, and when it did the challenge was met with great success and important results.

7

Europe

IN THIS CHAPTER the intention is to examine the progress of dam-building in European countries other than Spain between the Middle Ages and the nineteenth century, a period when Spanish dam-building flourished, as we have seen. Outside Spain varying conditions—political, geographic, climatic, economic—caused dam-building sometimes to develop along parallel lines, but not always. The simplest way to build up the picture is through a consideration of the achievements of individual countries—Italy, Austria, Germany, France and Great Britain.

Italy

In two senses the situation in Italy can be said to have been similar to that which prevailed in Spain. The southern tip of the mainland and Sicily have a very dry climate for many months of the year, and in addition the area experienced a Moslem occupation. It is fair to wonder therefore whether these two factors encouraged the construction of dams and irrigation works in southern Italy in the way that they did in southern Spain.

The answer in short is no. The climate, especially in Sicily, is dry enough to warrant irrigation, but the regime of the rivers and the terrain were not, it would seem, so conducive to the establishment of canal systems. Many of Sicily's rivers are dry for most of the year but then torrential for a few months in the winter. Of more importance, however, was the fact that the Moslems' occupation of Italy was short-lived and turbulent. Indeed, their presence on the mainland amounted only to a few raids in the ninth century. The Moslems' control of Sicily lasted a couple of centuries, from the middle of the ninth to the middle of the eleventh, and during this time, although the cultivation of such things as oranges, sugar and cotton was introduced, no large-scale irrigation schemes were built.

The development of dam-building in Italy is centred, in the main, on the north of the country. The achievements of the Romans have already been covered in Chapter 2. The biggest of the three dams at Subiaco survived until 1305, although its use as a source of supply to the Anio Novus ceased before that date. Furthermore the dam's collapse prevented the structure from exerting any influence on later developments as an example of what could be done.

Ever since its rise to prominence at the end of the twelfth century, Milan has been the centre of an area in which the construction of dams has been of great importance as part of a steady struggle to utilise and control the River Po and its various northern tributaries.

As early as 1179 work was begun on an irrigation canal from a point on the River Ticino near Oléggio to Milan. In addition to providing irrigation water, the canal was also used to maintain the water level in the moat around the city and to power a number of water-mills. In 1269 the canal was rendered navigable as an aid to the construction of Milan cathedral and took the name Naviglio Grande.[1]

For nearly 800 years the Naviglio Grande has been fed from substantially the same dam. It is a long low structure built obliquely across a bend in the Ticino, three miles east of Oléggio. Many a reconstruction has been necessary over the centuries, notably in 1705, 1819 and 1846, so that it is now utterly impossible to know what the weir was like originally. However, its present dimensions—920 feet long, about 6 feet high and varying in thickness from nearly 60 feet down to 30 feet—are perhaps representative of the medieval structure.

Another thirteenth-century dam is to be found at Cassano on the River Adda. It was built to feed the Muzza canal which irrigates land to the south-east of Milan. Nearly 1,000 feet long, the dam is built obliquely across the river, and numerous reconstructions, including a quite recent one, preclude any possibility of determining the structure's original form.

In the fourteenth century five more diversion dams were built for irrigation canals.[2] Three of these were on the River Sesia near Vercelli, one was on the River Ticino at Galliate, and the fifth on the River Dora Baltea at Rivarossa. In the fifteenth century these same three rivers, the Ticino, Sesia and Dora Baltea, which are the Po's principal tributaries to the west of Milan, were dammed for yet more irrigation canals. Among these was the Ivrea canal, built in 1468 and fed from a huge diversion dam on the southern edge of the town of the same name.

The Ivrea dam, substantially rebuilt in 1651, is about 1,500 feet long and diverts the waters of the Dora Baltea into a canal at its southern end. The dam raises the water level about 15 feet, and after 500 years of uninterrupted use is an impressive indication of what Italian dam-builders were capable of in the fifteenth century. By the year 1500, then, the region around Milan was equipped with a considerable network of canals whose total length was hundreds of miles.

Between 1462 and 1470 Milan was provided with its second navigable waterway, the Martesana canal. This was the work of Bertola da Novate, and included a dam and overflow weir at Trezzo just below the remains of Bernarbo Visconti's bridge. Recently the old dam has been replaced, but traces of it can still be detected. The dam itself was made of a wooden framework packed with large stone blocks and anchored to the river-bed on a piled foundation. Beyond the fact that the dam was 300 feet or so long, no other details have survived.[3]

Without doubt the most famous name which can be connected with the Martesana canal is Leonardo da Vinci's. It was at the Milan end that he built his famous pound-lock with mitre-gates. While there is nothing to suggest that Leonardo built any dams, it is known that he drew up plans for some. Near the end of the fifteenth century he addressed himself to the problem of protecting Florence from the ever-present threat of the River Arno's floods. His scheme involved the construction of reservoir dams above Florence in the Arezzo basin, where he intended to impound excess run-off and then release it gradually in periods of low flow.[4] In addition to using the dams for the flood-protection of Florence, Leonardo also worked on a scheme for a navigation canal between Florence and the sea. For this he was also dependent on the Arezzo dams as the means of regulating the Arno at all times.

Nothing came of this scheme, nor of a later one which Leonardo worked out. The Arno's floods, however, remained a serious threat to Florence, and during the sixteenth century more proposals to deal with the problem featured the construction of dams. In the case of a proposed dam across the River Bisenzio, the original drawings of the engineer, Gherardo Mechini, have survived.[5]

Apart from the often rebuilt irrigation dams around Milan, the oldest dam in Italy is near Cesena.[6] The Cento dam was built about 1450 by Domenico Malatesta, a late member of the family under whom Cesena flourished from 1379 until 1465.

The Cento dam (Pl. 39) stands across the River Savio about two and a half miles upriver from Cesena. It is 234 feet long at the crest and about 45 feet thick at its base. Recently the dam has been heightened by means of a parapet wall 4½ feet high which runs straight along the crest of the original structure. Another recent addition is a thick concrete facing on the long sloping air face. Clearly this has been provided to protect the dam from erosion; there is no separate spillway, and the Savio is allowed to pour freely over the top of the structure.

In spite of these modifications, however, the profile of the fifteenth-century dam can be deduced. Its vertical water face was some 19 feet high and the crest a few feet wide. Below the crest, the air face sloped downwards for a length of more than 40 feet and then dropped vertically about 5 feet to the level of the foundations. The structure was designed, then, as a straight gravity dam of more than ample proportions.

The Cento dam is unusual in that it is made of bricks and wood (discounting the concrete additions). Presumably these materials were the most readily available to the engineer of *c.* 1450, and although they are not ideal for a river dam without proper spillways they were obviously employed with success. It seems that the dam consists of bricks set in lime mortar, and the mass is bound together within a framework of wooden bars and oak poles. Very probably this wooden framework extends below the brickwork into the foundations.

Brick was used also to build the dam's abutments. At the right-hand end a high and massive brick wall supports the structure and also serves to guide overflow away from the river bank, thus preventing undermining of the abutment. At the other end the brick abutment is part of the old mill house which the dam once supplied with water.

Below the mill is a long canal carried along an embankment (also made of brick), and this channel once used to carry water to other water-powered mills further downstream. Nowadays the mills are no longer in use, but since the beginning of this century the dam has continued to be of service by providing a supply of water to two small hydro-electric plants near by.

Over the years the Cento dam has collected a good deal of silt, but since it is not required to store water this is not a serious problem.

Another early Italian dam was built about 1600 to create the Lago di Ternavasso[7] on a large and beautiful estate eighteen miles south-east of Turin. The dam is made of earth and stands to a maximum height of about 25 feet. Its crest thickness varies between 17 and 22 feet, while at the base its thickness may be as much as 50 or 60 feet. The embankment is so overgrown that precise measurements are impossible. The topography of the site precluded the construction of a straight dam, and the structure is therefore in the form of a huge Z. The three 'legs' of the dam are 690 feet, 230 feet and 150 feet in length, making a total of nearly 1,100 feet along the crest. The central section of the embankment, 230 feet long, is at right angles to the end sections.

The dam's water face is sealed with a vertical brick wall 2 feet thick, faced with a rough layer of mortar. So that the weight of the earth embankment will not push it into the empty reservoir, the brick wall is supported by a series of tapering buttresses, each one being 6 feet wide. This buttressing—it is clearly visible in Plate 40—is similar in function to that of the Proserpina dam in Spain.

The dam of Ternavasso is well equipped hydraulically. At the right-hand end there is an overflow spillway 12 feet wide and a foot or so deep which is adequate to deal with excess flow from the reservoir. The River Stellone is not often likely to produce severe overflow conditions. The dam has three outlet wells through which water can be drawn. One, the smallest, is near the spillway and the other two are installed in the longest leg of the embankment, one near the middle and the other near the dam's left-hand end. All three work on the same principle: each one consists of a semicircular wall which projects into the reservoir, thus forming a vertical well adjacent to the dam's brick face. A series of circular holes, 6 inches in diameter and 2 feet apart, allow water to run from the lake into each well, and the flow is controlled, very simply, by fitting or removing a set of large wooden bungs. At the base of each outlet well a horizontal gallery directs the outflow to canals below the dam.

Only the central outlet shaft is deep enough ever to have drawn a supply from the reservoir at very low water levels. Nowadays it cannot do this because of sediment. Thick beds of silt and debris have accumulated during three centuries or more of use, and a good deal of the reservoir's potential capacity of 300,000 cubic metres (240 acre-feet) has been lost. When the water level is more than about 15 feet below the dam's crest, the other two outlets are left high and dry. The dam has no device whereby the silt deposits can be flushed away.

The Lake of Ternavasso was formed to irrigate the lands of the estate on which it stands, a function which it still performs. It is also the home of hundreds if not thousands of waterfowl, and when they are disturbed, anyone standing on the dam's crest is subjected to the most ear-splitting racket imaginable.

As far as can be determined and excluding the Subiaco dams, the dam of Ternavasso is the only reservoir structure in Italy dating from before the year 1800. In the nineteenth century more were gradually added, especially for water supply. Earth gradually gave way to masonry for big dams, and near the end of the nineteenth century Italy's hydro-electric resources acted as a great stimulus to dam design and construction. In the twentieth century Italy has become a major force in the development of dams.

In carrying our story across the Dolomites from Italy to Austria it is essential to mention the most important and the most interesting old dam in Italy—a structure whose origins are not, strictly speaking, Italian at all. The Ponte Alto dam (Pl. 41) is on the River Fersina, a tributary of the Adige, about a mile and a half east of Trento. The city of Trento stands on the route to the highly strategic Brenner Pass, and in its long and turbulent history has been at various times under the control of Germany, Austria, France and now Italy. Rather than attempt to decide which of these countries can claim credit for building the dam, let us say that it is Tyrolean.

That the Ponte Alto dam is an amazing piece of work will be apparent. The present thin arch dam is not, however, the first structure to have been built on the site.[8]

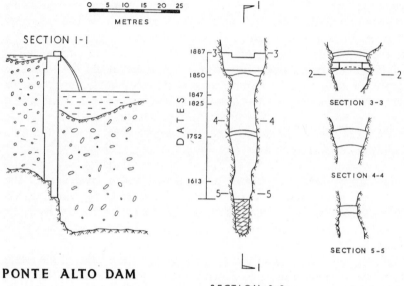

PONTE ALTO DAM

Figure 20   The Ponte Alto dam on the River Fersina above Trento.

The River Fersina, like dozens of other streams in the Dolomites, normally has a modest flow, but in the spring, when ice and snow melt in the mountains, it produces sudden and torrential floods charged with great quantities of rocks, trees and other debris. It was in order to protect Trento from these spates that the Fersina was dammed just above the city. As is obvious from Figure 20, the river at the chosen site flows through a deep and narrow gorge. A more perfect location for a dam would be hard to find, although it was to be some time before the gorge's potential was fully exploited.

The first dam was a small wooden affair constructed in 1537 by Franco Recamati of Verona. In 1542 it was destroyed by floods. In 1550, at the instigation of Prince Cristoforo Madruzzo, the dam was reconstructed in stone and cement and this time survived thirty-two seasons of use.

At the beginning of the seventeenth century the engineers appointed by Prince Carlo Madruzzo realised that the Ponte Alto site was ideal for a thin arch dam. Between 1611 and 1613 such a structure was built; it was 16 feet high, $6\frac{1}{2}$ feet thick and curved to a radius of $46\frac{1}{2}$ feet. It was made of cut masonry blocks without mortared joints, and for this reason was frequently damaged, notably in 1686.

In 1752 the dam was raised to a height of 56 feet, the addition to the dam being 13 feet thick at a radius of nearly 50 feet. The crest was fitted with a wooden overflow lip to prevent damage to the masonry. This timber spillway, however, was often damaged or destroyed by floods.

In 1824–5 the dam was raised yet again, this time to a height of 82 feet. The third section of the dam is curved to a radius of 47 feet and is $14\frac{1}{2}$ feet thick.

It is relevant to note that at the base of the dam's third stage there is a relieving arch, as Figure 20 shows. Probably, further extensions were anticipated at the time, and in 1847, and again in 1850, the fourth and fifth stages were duly added, bringing the dam's height up to $111\frac{1}{2}$ feet.

The final addition came in 1887, this portion being $14\frac{1}{2}$ feet thick and curved to a radius of $43\frac{1}{2}$ feet. The dam then stood, and still does, to a height of 124 feet. A relieving arch at the base of the 1887 extension perhaps indicates plans for further heightening, but this in fact has never been done.

The Ponte Alto dam was the first arch dam in Italy, and the portion of 1613 was the earliest arch dam in Europe. Its height/thickness ratio was high (between $2\frac{1}{2}$ and $8\frac{1}{2}$) at all stages of the dam's development, but in such a narrow gorge it was not, of course, necessary to use arches of great thickness. It is to the engineers' credit that they realised this. It seems that the dam features no mortared joints at all. All the masonry blocks are cut and fitted and, in the upper parts of the dam at least, are tied together with iron bars. At all levels the vertical jointing planes between the blocks lie along radial lines.

It is of interest that the various sections of the dam show subtle variations in thickness and radius, leading one to wonder whether the builders had some rule or formula with which they adjusted the radius and thickness of an arch of a given span. From the sets of figures quoted, however, no pattern is detectable, in which case the dam's dimensions presumably vary for no special reason. This, however, must have made for constructional difficulties, because each new arch configuration had to be matched to a lower one of slightly different shape. Another curious feature is the use of thicker arches at higher levels—the reverse of what is really required.

The dam's periodic heightenings were carried out in order to preserve its usefulness as a control dam. The structure was, after all, designed to store flood water and trap debris. As soon as it was full to the brim with rocks and silt it was unable to fulfil either purpose adequately and was therefore raised. The presence of relieving arches suggests that the inevitable cycle of events was recognised in the nineteenth century.

In 1882 the Ponte Alto dam was hit by a tremendous flood when another dam further up the River Fersina failed and released 1,200 acrefeet of water. The dam survived this ordeal, but it is significant that in 1883, when the final extension was begun, an unusual step was taken to prevent the structure ever being overloaded again. About 250 feet downstream another thin arch dam, 133 feet high, was built—it is called the Madruzza dam[9]—and this serves to impound water (and debris) to a height of 83 feet up the air face of the Ponte Alto dam. Thus today only the upper 41 feet of the Ponte Alto dam is, properly speaking, a dam; the rest is more or less equally loaded on both faces by water and sediment.

The reservoir above the Ponte Alto dam is nowadays full up with debris stretching more than a mile upstream. The air face of the dam is two-thirds covered by the presence of the Madruzza dam, while the gorge in which the dams stand is dark and gloomy, due to thick vegetation. One way and another the most interesting old dam in Italy has almost entirely disappeared from view.

The Madruzza dam, the one below the Ponte Alto dam, is one of several dozen arch dams built by Austrian engineers in the Dolomites for flood-control purposes in the second half of the nineteenth century, especially between 1880 and 1890.[10] Another is on the River Fersina about two miles above the Ponte Alto dam; a particularly fine one was built between 1884 and 1886 on the River Avisio near Lavis, and there are four more east of Rovereto on the River Terragnolo and one of its tributaries. They represent the first extensive use of arch dams in Europe although they are not for the most part very large. It is probable that the Ponte Alto dam served as the prototype from which they were developed.

Austria

Although Austria's contribution to dam-building was not an important one until the late nineteenth century, there are examples of Austrian dams several centuries earlier. The reason for their construction is rather unusual at first sight, but logical on reflection. Austria has always been a predominantly Roman Catholic country and is without a seaboard or easy access to the sea. In earlier times all over Europe Lent was much more scrupulously observed than it is now, and one of the requirements of the occasion was fish. Austria was obliged to catch her supplies in rivers and lakes, and of the latter many were artificial.

An impressively large example of one of these fish-rearing reservoirs was built in 1460 near Tarrenz, about thirteen miles north of Landeck in western Austria. Called the Spiegelfreudersee, it was one of Europe's first large earth dams.[11] Details of its construction do not seem to be known, but its principal dimensions can be given. It was 250 metres (820 ft) long, 8 metres (26 ft) high and about 15 metres (50 ft) thick at the crest. The base thickness without doubt was much greater. Information on other similar dams has not been forthcoming.

Germany

An early example of German dam-building occurred not far from the Spiegelfreudersee but at a much earlier date. Around A.D. 1000 the people of Augsburg were beginning to make use of water-power to drive mills in the town, and for this purpose they built the 'Lechwehr', that is to say a diversion dam across the River Lech. No dimensions or details of the dam are available.

Dams for water-power appear to have been the most numerous in the early history of German dam-building, and many were built as part of mining operations. Mining engineering has been a dominant theme in German technology, especially in the Middle Ages and the Renaissance. A major event in the development of German mining technology was the publication in 1556 of Giorgius Agricola's *De Re Metallica*.[12] For nearly two centuries this was *the* text on mining and metallurgy and it is one of the most important early books on technology to have survived intact, complete with its superb collection of drawings. *De Re Metallica* shows conclusively the extent to which German miners in the fifteenth and sixteenth centuries relied on water-powered machines of many types and for a whole range of operations. Agricola himself does not discuss dam-building as a factor in the use of water-power in mining; rather he assumes that a supply of some sort is available. However, it is clear that dams were a part of the mining scene in Germany.

One of the places where Agricola studied mining operations was the Oberharz, south of Brunswick; and it was here, in Agricola's time, that reservoir dams for mining work began to appear. Ultimately some sixty reservoirs were in use in the Oberharz, and of these one built between 1714 and 1721 required a dam of impressive dimensions, the first large dam ever built in Germany.[13] The Oderteich dam is 151 metres (495 ft) long, 22 metres (72 ft) high, 16 metres (52½ ft) thick at the crest and 44 metres (144 ft) thick at the base. It is of interesting construction. It consists of a central core wall of granitic sand sandwiched between two outer walls of granite blocks, roughly cut to shape and fitted together. The spaces between the blocks are packed with earth and moss. Although a crude example, the Oderteich dam appears to be the oldest rock-fill dam in Europe, and it was the first dam in Europe of any size built to create a hydro-power reservoir.

France

Ever since the Romans built one of Europe's first water-powered mills (for grinding corn) at Barbegal near Arles, France has been to the fore in the development of hydro-power machines: various types of water-wheel for many centuries and more recently the turbine. A factor in France's leading role in this development has been, of course, her considerable water-power resources in the form of numerous large rivers supplied from high mountain ranges that receive lots of rain and snow.

There is evidence of water-mills existing in France as early as A.D. 600 and by the eleventh and twelfth centuries the use of water-power extended rapidly and was applied to a whole range of technical operations apart from milling, the task for which it was most frequently employed in the early Middle Ages.

It was for water-wheels that dams were first built in France. In 1171 the Count of Toulouse gave the Bishop of Cavaillon the right to divert water from the River Durance in order to operate flour mills. In this case the water, once past the water-wheels, was used to meet local irrigation requirements by means of a canal system. In 1191 the River Drac was dammed near Grenoble by means of an earth embankment which collapsed in 1291. This is the first recorded failure of a dam in France. A few years earlier, however, another water-power dam was destroyed deliberately. William the Breton (*c.* 1159–*c.* 1224), in his account of the exploits of the French king Philip Augustus, describes how Philip besieged the town of Gournay near Beauvais and hastened its surrender by breaking the dam which fed water to the town's mills.

The city of Toulouse has had a long and continuous history of water-power usage. In the twelfth century the River Garonne was being used to drive floating mills, and then in the thirteenth century the river was dammed on the western side of the city in order to drive the first of the celebrated mills of Bazacle. For that time, a diversion dam 900 feet long was an impressive and expensive undertaking; the Garonne is a deep river with a considerable rate of flow all the year round and occasionally it is liable to produce destructive floods when the flow can be as much as 200,000 cusecs. Such floods have been a constant threat to the dam of Bazacle for 700 years, but nevertheless there has been one on the site for more or less the whole of this period.

The first Bazacle dam appears to have been a composite structure of earth, wood and stones. The Garonne must surely have played havoc with a diversion dam of this type. However, old prints of Toulouse show[14] that these materials continued to be used in various combinations right down to the eighteenth century. In 1709 an exceptional flood completely destroyed the dam, and an engineer called Abeille was assigned to rebuild the structure in a more permanent fashion.

The idea of a long oblique weir was abandoned, and the new structure was built straight across the river.[15] It was about 16 feet high and nearly 70 feet thick. Having a long sloping crest, this structure was much better equipped to resist the passage of the Garonne, and its stability was further enhanced by a massive piled foundation driven deep into the river-bed.

In 1722 and again in 1835 the dam successfully survived two of the most severe floods the Garonne has ever produced, and it is basically Abeille's dam which still spans the river at Bazacle. The most important recent addition has been a masonry facing.

Before 1800 hundreds of dams were built in France as part of the country's development of its hydro-power resources; the vast majority were very small. So, too, was an early French irrigation dam. For the most part France did not have an irrigation problem, although in parts of southern France irrigation offered definite advantages. As mentioned earlier, the Bishop of Cavaillon used his twelfth-century dam on the Durance for irrigation as well as milling. A later irrigation dam was built on the same river in 1554 by the Italian-born Adam de Craponne.

The so-called Craponne canal was in fact the first large waterway ever built in France, but unlike its famous successors was not intended for navigation. It was designed to carry Durance water to the dry, although fertile, lands of Provence in the Rhône valley. It was a sizeable piece of work, beginning at Cadenet on the river, winding its way across country to Lamanon and there dividing into two branches, one draining into the Rhône at Arles and the other into the Étang de Berre. There were several branches forming a total length of canals of more than 100 miles.

At Cadenet, de Craponne decided, quite deliberately, to build a temporary dam. At the intake of the canal, 492 feet above sea-level, the Durance bed consists of thick and shifting beds of sand and gravel across which it was judged to be impossible to erect a permanent dam. Instead he constructed a 'barrage volant', a cheap and unsophisticated dam of stakes, fascines and stones which was expected to last only a short time and whose frequent reconstruction was accepted as part of the system's year-to-year operation. Dams of this type have often been used in certain circumstances, and examples can still be seen in various places, but no longer now at Cadenet. The Craponne canal's headworks have recently been rearranged, and a few stakes in the river-bed are all that remains of the last diversion dam across the Durance.

The most extraordinary engineering work in seventeenth-century France was the famous Machine of Marly; it has often been written about. This outlandish mechanical contrivance was designed to supply water to the fountains in the gardens of the palace of Versailles. Basically it comprised fourteen large water-wheels which operated three sets of pumps, 221 separate pumps in all, and between them these were intended to lift water 530 feet from the Seine to an aqueduct four miles long. The 'machine' took four years to build (1681–5), cost a great deal of money, produced very little power for pumping and became famous for the fantastic amount of noise it generated.

The wheels were powered from a masonry dam across the River Seine for which no details have been uncovered. From various old engravings[16] of the installation, however, the dam is seen to have been a small affair which raised the Seine's level about eight feet. The Marly water-works is significant mainly in a negative sense. It demonstrated finally that water supplies could not be effectively provided by river dams driving sets of pumps. It was ultimately realised that the correct technique was either a high-pressure pumping system driven by steam engines, or a high-level reservoir feeding aqueducts, the latter being, as we have seen so often, an old-established concept. It is interesting, however, that for a time a certain passion for machinery blinded engineers to the advantages of reservoirs.

Leonardo da Vinci spent the last three years of his life (1516–19) in France under the patronage of Francis I. One of the last engineering projects to occupy him was that for a canal between the Atlantic and the Mediterranean. Of the two routes he considered, the more southerly one linking the rivers Garonne and Aude was essentially the one followed by the Canal du Midi or Languedoc canal 150 years later. Built to a large extent upon the experience gained from the Briare canal (completed in 1642) and a predecessor of the eighteenth-century canals of Bourgogne and Charolais, the Canal du Midi was the greatest civil engineering work of the seventeenth century, not just in France but in Europe. The canal, still fully operational, is 150 miles long, rises 206 feet from the Garonne at Toulouse to the summit, and then falls 620 feet to the Étang de Thau. There are a hundred locks, three large aqueduct bridges, a tunnel and numerous weirs, road-bridges and control works. Here was civil engineering in the grand manner and a triumph for two men—Jean Colbert, Louis XIV's chief minister, and Pierre-Paul Riquet, an engineer of great talent and dedication.

One of Riquet's main problems was the provision of a supply of water to the summit-level of his canal. Indeed, this was a basic problem in the construction of all early canals which traversed hilly or mountainous country and, as we shall gradually discover, had significant repercussions on dam-building.

To the north of the canal's summit, suitable sources of water were found in the rivers of the Montagne Noire, the only drawback being that more water than was needed was available in the winter, and less in the summer. Riquet therefore took the logical step of building a reservoir to store the surplus winter run-off for use during the following summer.

The dam of St-Ferréol[17] was built across the River Laudot about two miles south-east of Revel. At the place in question the river flows through a natural basin, and this was turned into a reservoir of 5,400 acre-feet capacity by damming it at its south-western end. The wide and shallow shape of the valley and the availability of suitable materials for construction persuaded Riquet to build an earth dam of impressive size. The dam of St-Ferréol (Pl. 43) is 2,560 feet long at its crest, has a maximum height of 105 feet above the river-bed and at the centre has a base thickness of more than 450 feet.

The earth fill which comprises the body of the dam is supported at its upstream and downstream faces by two thick masonry walls. The upstream wall is nearly 48 feet high and has an average thickness of 20 feet. The downstream wall is 60 feet high, nearly 30 feet thick at the base and 17 feet at the top. Half-way between these two walls is a masonry core wall which is 17 feet thick and extends to the full height of the dam; its top is in fact the crest of the dam. The three masonry walls run the full length of the dam at their respective positions in the profile, and all three are built firmly into bed-rock to a depth of 10–13 feet.

The spaces between the walls are filled with a mixture of earth and stones well beaten into place during construction. The upstream section of fill does not reach right up to the crest of the central core wall. Consequently when the reservoir is full the whole surface of the earth embankment is submerged, and so to prevent leakage it is covered with a layer of clay more than 6 feet thick. Any water which succeeds in penetrating this layer is retained by the masonry core wall.

The dam of St-Ferréol has a large low-level outlet tunnel which winds through the base of the dam along the line of the Laudot's original course. Where this gallery intersects the central core wall a set of outlet valves is installed. They are reached by means of a tunnel whose entrance is in the dam's downstream retaining wall. At its upstream end the low-level outlet tunnel is connected to a vertical well pierced by holes so that the impounded water can run into the shaft. This arrangement is reminiscent of the devices used in Spanish dams. An even more striking similarity is the use of the flushing technique to clear the reservoir of silt. Because it is so big, the low-level outlet tunnel meets this need itself. A large door at the base of the outlet well is opened up, as are all the other sluices within the dam, and the water-pressure does the rest.

So long as the reservoir is nearly full two high-level outlet sluices can be used to draw water. The higher of these is near the left-hand end of the dam, and near by is the overflow spillway. This was an essential part of the design because, had water been allowed to flow over the top of the dam, it would soon have cut into the earthen embankment below.

Work on the St-Ferréol dam was begun in November 1667 and completed four years later. The total volume of the construction is 208,000 cubic yards, of which 155,000 cubic yards are earth fill, and the whole of this was placed by hand. When Louis de Froidour visited the site during

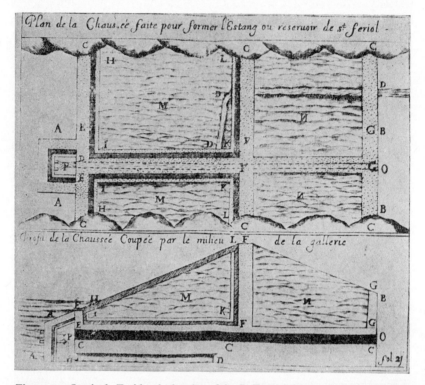

Figure 21   Louis de Froidour's drawing of the St-Ferréol dam made during his visit to the work during the course of construction. (Courtesy of the Institution of Civil Engineers.)

the course of construction he found hundreds of local workers toiling back and forth, each with a load of spoil. For each trip they were paid a penny.[18] Above is shown Froidour's drawing of the dam. It is not an accurate rendering of the structure in its final form, but is interesting nevertheless as an early technical drawing of a dam.

The dam of St-Ferréol, the first ever built to supply water to a summit-level canal, has served its intended purpose for close to 300 years and, so far as is known, without a hint of trouble. But only for a hundred years or so was it the only dam supplying the Canal du Midi. In April 1667

Pierre-Paul Riquet wrote in a letter to Colbert, 'I have all the water-supply that I require, and the invention of my reservoirs will furnish me during summer with sufficient to render the navigation perpetual.'

It is interesting that Riquet mentions 'reservoirs', and in fact it may be that at one stage there were plans for more than one. Among Riquet's engineers during the canal's construction was a young man called François Andreossy, born in Paris in 1633 of Italian parents. François Andreossy—he was the great-grandfather of the A. F. Andreossy to whom reference was made in Note 17—mentions that he had the idea of building a supply reservoir not on the Laudot but on the River Lampy, about eight miles to the east. Towards the end of the eighteenth century, as the Canal du Midi carried more and more traffic, the Lampy dam became a necessity. It was built between 1777 and 1781 and is still in use.[19]

The Lampy dam is quite unlike that of St-Ferréol. It is a masonry dam 385 feet long at the crest and 53 feet high. The crest thickness is 17 feet and both faces of the dam are steeply inclined. At a level $32\frac{1}{2}$ feet below the crest the thickness increases, by means of a horizontal sill on the water-face side, from $23\frac{1}{2}$ to $27\frac{1}{2}$ feet. Below this the dam broadens to a base thickness of 37 feet.

The whole structure stands on a masonry platform, 46 feet thick and $6\frac{1}{2}$ feet high, which is firmly anchored on its underside to the bed-rock. This platform projects several feet from the air face of the dam at its base, and is thereby able to support the ten buttresses which are distributed at 35-foot intervals along the dam. They are clearly visible in Plate 42. At crest level each buttress is $7\frac{1}{2}$ feet wide and projects downstream 7 feet. At their bases the buttresses widen to 13 feet but are only $6\frac{1}{2}$ feet long.

The Lampy dam was Europe's second buttress dam and was begun thirty years after the dam of Almendralejo. Conceivably M. Garipuy, the Lampy dam's designer, knew of the Spanish structure. Certain features of the St-Ferréol dam suggest that engineers in south-western France may have been familiar with Spanish practice. It cannot be said, however, that at Lampy the buttressing principle is well developed; the buttresses are far too slender to contribute much to the dam's stability. In fact the Lampy dam has not been a wholly satisfactory structure.

Writing in 1804, A. F. Andreossy mentioned that quantities of lime had been thrown into the reservoir close to the dam's water face in the hope that the lime would percolate into the body of the structure and thus seal cracks through which water was leaking. He regarded this as a highly ingenious idea and even pronounced it a success. But in the long run it was not the answer, and today the dam's air face is leaking considerably.

Even more alarming are dozens of large nuts projecting from the dam's crest. These are screwed to the tops of long iron bars which at some time have been inserted right through the dam into its foundations and then tensioned from the top. Evidently it is feared that the dam has insufficient weight and too little buttressing to resist the water-pressure unaided.

The Lampy dam has three outlets at different levels. Each one is opened and closed by means of a sliding sluice-gate on the water face of the dam. The outlet tunnel near the crest is dated 1780, and close to it is the spillway.

One other canal reservoir was built in France in the eighteenth century. In 1766 the Caromb dam (Pl. 44) was constructed across the River Lauron for the canal of Carpentras. It is a masonry dam, 207 feet long, 69 feet high and some 25 feet thick at the crest, a gravity structure of ample proportions and simple equipment. Training walls guide overflow over the crest and there is a single low-level outlet sluice.

Large summit-level canals were the finest French civil engineering achievements of the seventeenth and eighteenth centuries. It was in order to supply them with water that big dams were first built in France. The other stimulus to dam-building in France, but on a much smaller scale, was water-power. In Great Britain it was the same two factors—navigation and water-power—which encouraged the construction of dams. But before the last decade of the eighteenth century Great Britain had not a single dam of any size. Nevertheless the early British record is not without its interesting aspects.

Great Britain

Considering its great age, England's first recorded dam is remarkably well documented.[20] Seven hundred years ago the Winchester–Alresford area in Hampshire was prosperous because of its wool and cloth industries. In 1189 Godfrey de Lucy, Bishop of Winchester, decided to improve the commercial potential of Winchester and Alresford by making the River Itchen navigable along its entire length between Alresford and Southampton Water. This work required the construction of a series of flash-locks or stanches whose subsequent operation depended on a supply of water in excess of the river's natural flow. Bishop de Lucy therefore constructed a reservoir near Alresford; it can still be seen (Pl. 45) although it no longer serves the long-since-abandoned waterway.

The dam stores the water of two small streams to the north of the town. The earth embankment is nearly 250 feet long, 60 feet thick at its base and 30 feet at its crest, and rises to a maximum height of 20 feet. For the time, around A.D. 1200, this was a sizeable piece of construction work. Unfortunately the dam's neglected condition and great age preclude any possibility of determining how it was built, how it was faced or what sort of hydraulic equipment it may once have utilised.

The reservoir's capacity is difficult to work out, but when full it would have covered about 200 acres. This is no longer the case. The formation of swallow holes now limits the reservoir's extent to about 60 acres.

Even after the demise of the navigation, the Alresford dam was maintained for the benefit of the town's mills. Today it also serves to carry a road along its crest.

Domesday Book of 1086 contains a well-known technological statistic: that in eleventh-century England there were 5,624 water-mills. It is inconceivable that at least a few of these were not powered by dams of some sort, small though they must have been. Subsequently, water-wheels were installed all over England.[21] They were used for all manner of operations—corn-grinding, fulling, iron-working, mining and water supply—and very many involved the construction of small dams. The favoured materials of construction were the obvious ones—earth, wood, stone and brick. Active in the medieval development of England's hydro-power resources were monastic communities. Water-mills were an integral part of virtually every monastery which stood near a suitable water supply. At Fountains Abbey in Yorkshire the original layout of the water-power system still survives, although the present diversion dam across the River Skell is not the original one.

Godfrey de Lucy's river navigation in Hampshire was the prototype of subsequent similar schemes.[22] Whenever a river began to be used for navigation as well as water-power, the dams always became the centre of argument and litigation, and not infrequently spontaneous fights broke out. While the millers insisted that the head of water above a dam was essential for the operation of their water-wheels, which was true, the navigators claimed the right to lower the water level by operating the flash-locks which allowed them to pass the dams. In the north of England and Scotland mill-dams also fell foul of the interests of fishermen. In 1683, for instance, a group of Scottish gentlemen planned a raid on the newly completed weir across the Tweed at Berwick, but were thwarted by the local people.

In the sixteenth century English river navigation techniques were given a fresh slant with the construction of the Exeter canal.[23] In 1313 navigation of the River Exe between the city and sea had been ruined by the construction of Countess Weir, a mill-dam which has now disappeared. For 200 years this dam and others were in continuous use on the river, and eventually even their removal would not have rendered the Exe navigable, so much had the river bed been altered. In 1566 the problem was solved by John Trew who cut the Exeter canal, nearly three miles in length, along the river's south-western bank.

The Exeter canal was the first in England to utilise pound-locks and was supplied with water by 'Trew's Weir', a masonry dam built across the

Exe just below the head of the canal. This dam, within a hundred years, was the centre of the traditional quarrel between navigators and millers when a certain George Browning began to utilise the dam to supply water to his fulling mills. An indirect result of these squabbles was the construction, in 1676, of a new dam very close to the site of Trew's Weir; it is this dam (Pl. 46) which still allows the occasional use of Exeter's pioneer waterway.

The technique used to make the Exe navigable became standard practice on rivers all over England. In the seventeenth and eighteenth centuries rivers such as the Thames, Wey, Kennet, Aire and Calder were all equipped with sets of dams and locks as part of the construction of the lateral cuts which markedly improved England's internal communications; and from them ultimately developed the canal systems of the second half of the eighteenth century, the works of men such as Brindley, Smeaton, Rennie and Telford.

More than anyone else John Smeaton established civil engineering as a profession in its own right in Great Britain. What makes his career particularly interesting is the survival of his notebooks compiled during the course of his employment on a very varied selection of projects, including a number of dams.

There is not space here to deal with everything that John Smeaton had to say about, and to do with, dams. In any case, much of it is detailed and tedious; a number of his dam projects were, for various reasons, never realised; and sometimes his reports and designs are to do with repairs and alterations to existing dams, as in the case of the Aire and Calder navigation, for instance. Three dams, however, all for water-power installations, are worth closer examination.

The Carron Ironworks[24] near Falkirk was opened in 1760. Smeaton was first introduced to its founder, John Roebuck, in 1766, and at various times during the next twenty years was often called upon for his advice and ideas.

From the outset the blowing engines and boring mills at Carron were driven by water-wheels, the required supply of water being taken from a diversion dam at Larbert on the River Carron and stored in a reservoir close to the works. Demand rapidly outstripped supply, and in 1767 the works had to be closed for three months due to a lack of water. Smeaton was called in to engineer a solution.

He proposed in the first place to enlarge the Larbert dam so that it would divert more water into the canal leading to the works' reservoir. The dam stands to a height of about 8 feet at a point where the river's width is 100 feet or so. The structure is curved in plan, has a vertical water face and a long sloping air face, down which the overflow runs. The canal head is just above the dam at its right-hand end.

Smeaton's second proposal was for another dam, further upstream at Dunipace Bridge, to supplement the role of the Larbert dam.[25] The designs for this structure were prepared in 1773 and reveal a dam some 200 feet long, 45 feet thick at the base and 5 feet high, with gently sloping air and water faces. Smeaton's drawing shows the dam as having a rubble

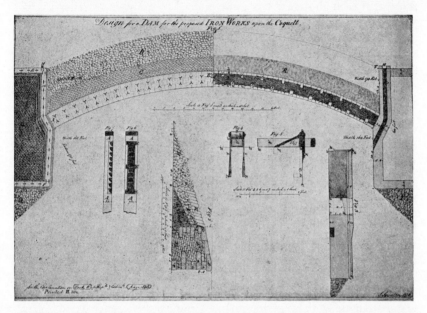

Figure 22   Smeaton's own drawing of the dam on the Coquet. Comparison of this with Plate 47 will show how little the dam has been changed in two centuries. (Courtesy of The Royal Society.)

masonry core faced with large flagstones laid on their edges. Two rows of sheet piling—one near the dam's upstream edge and the other at the tail of the air face—are specified as indispensable in any dam not founded directly on rock. The Dunipace dam can still be seen very much as Smeaton planned it, but it is not now in use.

In 1776 Smeaton was engaged to design an ironworks on the River Coquet in County Durham. As usual this was a water-powered installation for which a dam was required. It is shown in Plate 47 and today still stands exactly as John Smeaton designed it; his original drawings are shown in Figure 22 above.

The dam is curved in plan to a radius of 170 feet at the air face, which is vertical. The main wall of the dam is 8 feet high at the air face and 8 feet thick at the base. It consists of a rubble masonry core faced all over with large masonry blocks. The crest of this wall slopes upstream, and its water face is sealed with a wedge of earth near the base. This earthen seal is covered with a large quantity of rubble masonry protected against scour with flagstones laid on their edges.

The dam is supported at each end by massive masonry abutments, and Smeaton in his always meticulous fashion says that these were to be built first, in the middle of April, when the river was at its lowest. He further states that each abutment was to be pierced by a low-level gallery, 2 feet square, through which the river could be temporarily diverted. To achieve this he suggests that the leading edge of the dam's rubble masonry sealing wall be built to a height of $2\frac{1}{2}$ feet, thus forcing the river to flow through the already constructed galleries in the abutments. This is the earliest example so far located of specific instructions on what to do with a river in order to lay dry a dam's foundations.

The Coquet dam's survival over nearly 200 years is testimony to Smeaton's excellent design and the contractor's sound workmanship. Not that the dam's builder was left in any doubt as to how to proceed; John Smeaton overlooked nothing in his detailed instructions.[26] He prescribes the precise order to be followed in arranging the various components of the main dam and its protecting wall, how they should be placed, what to do in case of a sudden rise in the river's level which would flood the unfinished structure, and so on. Furthermore he lists the ingredients necessary to make a good mortar, and how to mix them. Included is 'Civita Vecchia pozzelana', in order to render the mortar 'hydraulic'. This is a reminder that it was John Smeaton who, while studying the reconstruction of the Eddystone Lighthouse, carried out the first systematic work on hydraulic limes and materials, such as pozzolana, which would impart hydraulic properties to ordinary limes.

While the whole of Smeaton's report on the Coquet dam cannot be quoted for want of space, his opening remarks merit inclusion for what they reveal of his attitude to the problem of dam-building in general.[27]

There is not a more difficult or hazardous piece of work within the compass of civil engineery than the establishment of a high dam upon a rapid river that is liable to great and sudden floods, and such I esteem the river Coquett, and such the dam here proposed to be erected; and when it is considered, that the performance of every part of the intended works depends upon the firmness and well-establishment of the dam; and further considered what loss, disappointment, trouble, and vexation, will attend a failure thereof, especially in the winter season, when such a misfortune is more likely to happen than at any other time,—it will readily be granted that too much care and circumspection cannot be used in putting the design here proposed into execution.

Summarising this chapter briefly, we find that dams for water-power were characteristic of most European countries over a long period. Only rarely were they large dams.

Flood-control dams were unique to Italy because of the special conditions prevailing in the Tyrol, and most of these were built by Austrian engineers. The Ponte Alto dam is the only early European arch dam outside Spain.

Irrigation dams are found only in Mediterranean regions and not in northern Europe.

England and France built many dams for inland navigation, but whereas in England the dams were small and associated with the navigation of rivers and the establishment of river navigations (i.e. rivers with lateral cuts), French engineers in the seventeenth and eighteenth centuries went in for wholly artificial summit-level canals which required the construction of the first big European reservoir dams outside Spain.

8

The First Half of the
Nineteenth Century

BROADLY SPEAKING, THE initiative in dam-building was now with the developing industrial countries. By 1800 the Industrial Revolution was well advanced in Great Britain, and during the nineteenth century industrialised societies quickly evolved in other countries, principally in Europe and in North America.

The causes of the Industrial Revolution and the ways in which its effects came to transform the world were complex. That it led to profound social and economic changes and set in motion a new era in human history is too well known to require discussion or elaboration here. What will emerge in this and subsequent chapters is the fact that dams played their part in the opening phases of the Industrial Revolution, that nineteenth-century social and economic conditions produced fresh problems which dams helped to solve, and that gradually the development of new technologies provided the opportunity for dams to fill new roles as well as continuing to meet traditional requirements on an ever-increasing scale.

A canal system was, of itself, not enough to stimulate an industrial revolution. France, after all, pioneered large inland waterways in the seventeenth and eighteenth centuries, yet no startling development of engineering or industry resulted. But in Britain, in association with other important factors, canals were undoubtedly contributory to the country's industrial development and rapid rise to world domination.

Initially canal engineers such as James Brindley concentrated on long contour canals and thereby avoided the expense of constructing locks and the problem of providing summit-level water supplies. But as the canal network expanded and in order to establish the integrated system which commercial interests logically required, canals had to be built over high ground. It was in order to feed water to these summit-level canals that England, in the last decade of the eighteenth century, constructed her first big dams. Some of them, very nearly in their original condition, are still in use, not necessarily for supplying canals, and a selection are worthy of our attention.[1]

The Peak Forest canal was built between 1794 and 1800 by Benjamin Outram. Its most famous feature is the Marple aqueduct bridge across the River Goyt, but it also has two reservoirs, both in the vicinity of Whaley Bridge. They are reported to have been built in 1794, but that they could both have been completed within the year that the project began seems doubtful.

The Todd Brook dam, just to the west of Whaley Bridge, has the greatest height of all the late eighteenth-century English canal dams—70 feet at the centre. Its crest, 7 feet wide, is about 700 feet long. The dam consists of a huge earth embankment (Pl. 48) whose faces slope at about 45 degrees, and the structure's thickness at the base must be of the order of 200 feet. There is a spillway at the right-hand end, and two low-level outlets through which water can be drawn to supply the feeders connecting with the canal not far below the dam. One of the outlets is equipped with escapes so that excess water can be released into the spillway channel. The latter is so arranged that it runs parallel to the crest of the dam and discharges at its lower end into the river.

Coombs dam, two miles south-east of Whaley Bridge, impounds the canal's other water supply. It is another earth bank, 52 feet high, 1,000 feet long and slightly curved. It has a single low-level outlet to feed the canal and two overflows. One of these is built into the crest at one end of the dam; the other comprises a vertical well set into the sloping water face near the centre. Unusually, both overflow levels can be adjusted by means of sluices.

The Huddersfield canal, begun in 1794, was long delayed in its completion (1811) due to the difficulty of driving Standedge tunnel; it is the longest canal tunnel in Great Britain and the one at the highest summit-level. The canal's water supply is taken from two reservoirs, one at Slaithwaite and the other very near the tunnel's eastern end at Marsden. Both are curved earth banks, between 55 and 60 feet high and more than 500 feet long. Their combined capacity is less than 450 acre-feet—enough, presumably, to ensure all-the-year-round operation of the canal.

The four dams so far mentioned were constructed to the same pattern. All are earthen embankments with sloping faces, the water faces being covered with a lining of masonry blocks to resist wave action and prevent erosion. Earth was used to build the dams largely because this material was most readily available and also because foundation conditions in many parts of England—soft sedimentary Mesozoic strata—favour earth dams to the exclusion of other types.[2] Puddled clay was employed to make the dams watertight, and this material quickly established itself as the standard one for the core walls of earth dams. In the case of another early canal dam, however—that built on the River Tamar in 1805 to supply the Bude canal—it appears that the core wall is of brick.[3]

All the dams are adequately equipped with low-level outlets and over-flow spillways, the latter, of course, being crucial in earth dams. Of much greater interest from the hydraulic point of view is the technique used to ensure that the rivers below the dams were never deprived of water. In other countries, in Europe that is, it was usual to allow all available water to enter a reservoir and subsequently to regulate the outflow at the dam. In England a different practice developed. At the heads of the reservoirs small weirs were built to divide the inflow into two streams. One of these is led round the reservoir in a specially built channel which returns to the river bed below the dam. This channel takes precedence over the other stream—the one which feeds the reservoir. Only when the flow of the river reaches a certain quantity does the regulating weir begin to discharge and thus allow the reservoir to fill. In this way compensation water is always supplied automatically. From the outset the construction of reservoir dams in England was treated with concern and suspicion by mill-owners and riparian proprietors, who feared that their livelihoods would suffer if the flow of rivers below dams was reduced or removed altogether. Their case was, of course, entirely valid, and there was many a legal action between them and the projectors of reservoirs in order to work out an equitable way of dividing up the available supply of water.[4]

The type of reservoir dam evolved in the 1790s set the pattern for all subsequent canal works and, later on, for many water-supply schemes. It would be tedious to deal with very many, but a few deserve to be singled out.

The Leeds and Liverpool canal, forty-six years in the making (1770–1816), is the oldest and most northerly trans-Pennine waterway and abounds in interesting civil engineering works. Two of the supply reservoirs, one above the other, are at Foulridge, adjacent to a tunnel nearly a mile in length. The upper dam is the more impressive work, an earth bank some 300 feet long and 40 feet high. At Barrowford is a third reservoir. It is an interesting structure because it is not built across a valley in the usual way. Instead, a hillside forms one side of the lake and the other three sides are all formed by the dam wall itself. There are other examples of this technique in the north of England, often as part of the water-supply arrangements to lead mines. Two can be found near

Nenthead, one of which, Perry's dam, is of considerable size, with an embankment half a mile long.

A number of famous engineers were involved in the construction of early canal dams. John Rennie, for example, was responsible in 1797 for the Rudyard dam, whose construction became essential when the Trent and Mersey canal began to carry so much traffic that Brindley's original underground water-supply system, using the drainage from a mine, was unable to cope.

The Trent and Mersey canal was provided with another reservoir in the late 1820s by Thomas Telford.[5] This was impounded by a dam at Knipersley, five miles north of Stoke-on-Trent, and its construction caused a good deal of difficulty due to leakage. Even when it was put to use in 1829 it was still less than watertight, although the trouble was ultimately cured. Two more dams designed by Telford were for the Birmingham and Liverpool canal, the last important project with which he was involved and the last major English canal before the railways swept all before them. The dams of Knighton and Belvide are therefore among the last of their kind in England, replicas of those built a mere forty years earlier.

The survival of many early English canal dams is proof of their sound design and construction, adequate dimensions and proper provision of spillways and outlet works. But success did not attend every venture. The Leicester Navigation was intended to take its water supply from Black-brook reservoir on the edge of Charnwood forest five miles west of Loughborough.[6] The dam was begun in April 1795, and in February 1799 heavy rain and melting snow combined to produce severe flooding which overloaded the dam's spillway and cut a great hole in the dam itself. By 1801 the dam had been rebuilt, but already the Navigation was proving to be a commercial failure. When the dam failed again, the waterway's fate was sealed.

It is curious that the nineteenth-century earth dam's successor, a concrete gravity dam built in 1906 to supply Loughborough with water, has had its troubles too. In 1957 it achieved the distinction—unusual for an English dam—of being damaged, not seriously, by an earthquake which cracked and dislodged a part of the parapet wall.[7]

The canal engineers brought big dams on to the British engineering scene, adding yet another dimension to the civil engineering picture in growing industrial Britain. Even places far removed from the principal industrial centres were concerned in these works. Chard, for instance, was at one time involved in a plan to construct a canal from the Bristol Channel to the English Channel; all that was ever built, however, was a short branch to the Bridgwater and Taunton canal. Nevertheless a reservoir was constructed to feed the canal and, judging from the following extract from the Chard Union Gazette of 5 April 1841, it made something of an impact in the locality.[8]

> The reservoir has risen considerably during the past week and continues to present new features of interest and attraction as it is viewed from various points in the neighbourhood—its broad expanse of surface of the deepest blue, resting sometimes in peaceful repose, and at others ruffled . . . and tossed to and fro, covered with crested waves of no mean size, and its shores washed by breakers, and besprinkled with angry foam.

The writer was also anxious to find

> a name for this fine sheet of water, one which shall be in classically good taste, consistent with the nature of things (as the school-men say) and yet perfectly intelligible to, and adapted to the articulating organs of the *profanum vulgus*.

In the first half of the nineteenth century canal dams were by no means unique to England. France, the country which had pioneered them, continued the tradition with a number of big structures.

The Couzon dam was in fact begun before 1800; its construction spanned the twenty-three years between 1788 and 1811. It is surprising that the structure has been so ignored by historians, because not only is it a big dam for the period but also it is a replica of Riquet's dam at St-Ferréol. The Couzon dam—it is situated some four miles south of Rive-de-Gier, near Saint-Étienne—is the only one in which Riquet's ideas were copied exactly.

The central masonry core wall of the Couzon dam is 200 metres (656 ft) long and 33 metres (108 ft) high. At the crest it is 4·9 metres (16 ft) thick and at the base 6·82 metres (22½ ft). The downstream earth embankment is 66 metres (216 ft) thick at the base and slopes up to the level of the top of the core wall. The dam's downstream extremity is a low masonry wall, 17 feet thick. The upstream earth bank reaches 52 metres (170 ft) into the reservoir and terminates in a thin masonry wall, 35 feet high. So long as the reservoir is at least half full, the upstream bank is completely submerged.

Like its forerunner on the Canal du Midi, the Couzon dam has two tunnels running right through the embankment at the deepest point. The lower of the two follows the original course of the River Couzon and is used to desilt the reservoir. Just above this scouring gallery is situated the outlet tunnel with its valves located at the water-face end. The dam is equipped with two spillways whose combined length is 40 metres (131 ft).

The Couzon dam was built in the first place to feed water to the Canal de Givors. When this waterway went out of use the dam was taken over by Rive-de-Gier for water supply, and this continues to be its function. Compared with the dam of St-Ferréol, the Couzon dam was not so well built. For many years it leaked badly and slowly slid downstream. Between 1895 and 1896 it was substantially rebuilt and ever since appears to have retained its 1,180 acre-feet of water without trouble.

Between 1833 and 1838 a canal was built between Nantes and Brest, supplied from three reservoirs, each one formed behind a masonry dam. The dam of Vioreau has one of the simplest profiles of any we have come across; it is rectangular, 11 metres (36 ft) high and 7·5 metres (24½ ft) thick, with a small sloping step on the crest.

The dam of Bosmeleac is also a gravity structure, 15·3 metres (50 ft) high, 4·3 metres (14 ft) thick at the crest and 8·5 metres (28 ft) thick at the base. Its water face is vertical, and the air face slopes. The profile is, in principle, more logical than that of Vioreau but dangerously slender all the same.

The last structure of the trio is the Glomel dam, 12·1 metres (40 ft) high, but this time it is the air face which is vertical while the water face is stepped. The dam's thickness increases from 0·6 metres (2 ft) at the parapet wall to 7·5 metres (24½ ft) at the base.

The three dams on the Nantes–Brest canal indicate clearly how little early nineteenth-century engineers, even in France, understood of the behaviour of gravity dams. The three structures, built for the same scheme and in the same short period of five years, show no conformity of cross-sectional shape whatsoever; in fact, if anything they suggest a divergence of opinion.

The confusion is further emphasised by a contemporary structure, the Grosbois dam, built between 1831 and 1837 to supply the Canal de Bourgogne, in origin a seventeenth-century scheme which was eventually begun in 1775 and finally linked the rivers Seine and Saône in 1832. The Grosbois dam[9] is 22·3 metres (73 ft) high and 550 metres (1,805 ft) long —a structure of some size. Its air face is nearly vertical and the water face is regularly stepped in six large increments, thereby increasing the dam's thickness from 6·5 metres (21½ ft) at the crest to 16 metres (52½ ft) at the foundations.

Despite its illogical shape, this dam would probably have performed satisfactorily had not the foundations proved to be a source of endless trouble. The structure was built on Lias clay, and when it was first required to hold water considerable leakage occurred. These leaks were sealed with masonry, and in 1837 the finished structure was filled up. More leaks occurred but, even more alarming, the dam was found to have slid on its clay base about two inches downstream.

Consequently in 1842 seven large buttresses, 8 metres (26¼ ft) long at the top, 11·8 metres (38½ ft) at the base and with a mean width of 10 metres (33 ft), were built against the air face, and in 1854 two more were added because the dam was still leaking and still sliding. For the rest of the century the dam continued to move slightly and, following the failure in 1895 of the Bouzey dam (see Chapter 9), an event which threw French civil engineers into something of a panic, it was subjected to remedial measures.

About 800 feet downstream an earth dam was constructed (in 1905) which forms a *contre-réservoir*, that is, a small reservoir impounding water to a height of 15·35 metres (50 ft) up the air face of the masonry dam and thereby relieving the latter of a good deal of pressure. It should be added that the earth dam then proceeded to experience serious problems of its own, and the strange Grosbois 'twin-dam' was only finally made safe in 1948.

Two other aspects of the Grosbois masonry dam are of interest. Because of its tendency to slide, it was subjected in its early years to careful measurement, and it was found that successive filling and emptying caused elastic deformations. This appears to be the first occasion on which engineers realised that masonry dams behaved elastically, and it is significant that quantitative studies of this type began in France.

The other point concerns the work of Alexandre Collin. It was at Grosbois as a young engineer that Collin observed, and measured, a large slip surface in the dam's clay foundation trench. Subsequently Collin was able to examine another example of this phenomenon, when the Cercey dam (also on the Canal de Bourgogne and completed in 1835) experienced a large slip on its upstream face. Collin proceeded to examine as many more cases of the instability of clay slopes as he could find and, as Professor A. W. Skempton has shown,[10] his work could have been the basis for a proper understanding of the behaviour of cohesive soils. In fact his valuable research went unnoticed for decades; had things been otherwise, it is conceivable that soil mechanics would have progressed far enough by the end of the nineteenth century to have improved the earth dam's poor safety record.

Just as canal dams were constructed in Europe, so they were in North America. It was in the state of Pennsylvania that two of the earliest high dams in the United States were built, both to supply canals. The Swatara dam on the Union canal was constructed before 1827 and stood to a height of 45 feet; details are not available. A much bigger effort was the South Fork dam near Johnstown, built in 1839 and abandoned as a canal supply only eighteen years later. It was a big dam, 840 feet long, 72 feet high and varying in thickness from 10 feet at the crest to 200 feet at the base. The dam had a core wall of slate, upstream of which was earth fill covered with a loose stone water face. The downstream half of the

structure was more of the rock-fill than earth-fill type; it consisted of a
pile of rubble masonry without a facing. The low-level outlet was a
masonry-lined tunnel supplied by means of a sluice gate at its upstream
end, the sluice being controlled from the top of a masonry tower standing
in the reservoir.

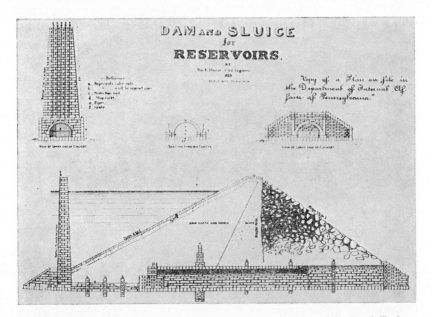

Figure 23   These nineteenth-century drawings show the layout of the South Fork or
Johnstown dam of 1839. Its collapse in 1889 is said to have killed 2,000 people and is
one of the most serious dam disasters on record. (From *Transactions*, ASCE, Vol. 24.)

Although it reflects the entirely random approach to the design of
earth and rock-fill dams which characterises early structures of this type,
the South Fork dam was well made and performed without serious
trouble for the canal operators. Subsequently it was used for fishing; then
between 1862 and 1872 it was employed by the Pennsylvania Railroad to
top-up locomotive boilers, and in 1875 it was taken over by a Pittsburgh
hunting and fishing club who raised the dam to a height of 75 feet.

The South Fork dam made its notorious mark on the history of dams
in May 1889.[11] Unprecedentedly heavy rain quickly filled the reservoir
and overloaded the inadequate spillway, which in any case the fishermen
had partially screened off. Overflow soon cut into the embankment, and
the whole structure collapsed; the water level is reported to have dropped
over 60 feet in a matter of minutes. The people of Johnstown further
down the valley had no chance of escaping the wave of water. It is said

G

to have killed about 2,000 of them, the effects of the flood being compounded by fires which broke out.

While the South Fork dam ultimately produced the first serious dam disaster in the United States, another early North American canal dam was, by contrast, a great success.

The Rideau canal, linking Ottawa with Kingston, was constructed between 1826 and 1832 by a team of military engineers led by Lieutenant-Colonel John By. In all, the waterway required the construction of fifty-two dams, some of which were simple earth embankments and not very large, while a few were masonry structures of some size. The biggest and most notable was the dam at Jones Falls about thirty-five miles northwest of Kingston.[12]

It stands at the head of a rocky gorge connecting two lakes, a point at which the canal is obliged to rise nearly 60 feet through a set of four locks. The dam consists of two parts. On the air face side is a masonry wall $62\frac{1}{2}$ feet high, $27\frac{1}{2}$ feet thick at the base and $21\frac{1}{2}$ feet at the crest. The length at the crest is 350 feet and the wall is curved to a mean radius of about 175 feet. Here then was the first arched dam in North America, and at the time it was built (1832) also the highest dam.

On its upstream side the masonry dam is backed up to its full height by an earth embankment, $127\frac{1}{2}$ feet thick at its base. Between the arched dam and the earth wall it was originally intended to place a puddled clay lining, but in fact most of the lining is of broken stone and mortar.

In order to build the dam it was necessary to control the river pouring through the narrow gorge. So, at its head, the gorge was blocked with a temporary dam fitted with wooden channels, one down each side of the dam site. The masonry dam and its earth wall were then built up around the channels, and these were finally both closed as soon as the dam had reached the height of its own overflow spillway.

The Jones Falls dam is today in perfect condition and still serves the function for which it was designed. It also supplies a head of water to a small hydro-electric station near by. It was while the intake of the penstock for the latter was being installed that the excellence of the dam's construction was revealed, and indeed it says much for John By and his men that they built one of the most splendid of early North American dams in difficult and demoralising conditions.

Perhaps the most significant sociological consequence of the Industrial Revolution was the growth of large cities. The migration of people to them was made attractive because it was there that work was to be found; it was made necessary because the needs of manufacture, commerce and industry had to be centralised; and it was made possible by the establishment of new and improved modes of transport: canals, ships, railways and roads. At the same time, however, this growth led to some of the most unpleasant and sordid aspects of nineteenth-century life. One particular side of the problem of large cities, which rapidly asserted itself, was that of water supply, not only for people but also for industry.

For centuries water supplies had depended, in the main, on the direct use of rivers, springs and wells.[13] Occasionally these sources of water were utilised in connection with pumping installations, such as those at Marly, London Bridge and Toledo, and not infrequently with crude systems of distribution pipes. By the nineteenth century such arrangements were found to be grossly inadequate. They yielded water which was insufficient in quantity or heavily polluted, or both. Essentially, the problem facing the nineteenth-century water-supply engineer was that of reintroducing the Roman type of system whereby clean and adequate quantities were intercepted wherever they could be found and then channelled to cities. Indeed, when viewed from a nineteenth-century standpoint, the Roman achievement in water supply must have seemed all the more remarkable.

The type of earth dam built in Britain for canal reservoirs was very much the one adopted for water-supply works. This was logical. The aim, after all, was basically the same even though the ultimate destination of the water was different. The earliest example of a water-supply dam appears to be the one built at Whinhill in 1796 for the supply of Greenock. For forty years it had the most undistinguished career: it failed in 1815, was rebuilt in 1821, failed again in 1835, and was later rebuilt a second time. As it is today, the structure is a typical earth dam, 800 feet long, 40 feet high, with a masonry lining on its gently sloped water face.

Scotland was very much to the fore in the construction of water-supply systems in the early nineteenth century, especially around Edinburgh. Even earlier, in the eighteenth century, John Smeaton had recommended the use of springs in the Pentland hills to the south of the city as a source of supply, and his views were subsequently corroborated by both Thomas Telford and John Rennie. Following a severe shortage of water in 1810, Telford suggested how the water of Crawley Spring might best be utilised and astutely foresaw that millers and landowners would object to its diversion. He proposed to solve this problem with a reservoir on Glencorse Burn from which compensation water could be directed down the valley. Even John Rennie's support for this project had no immediate effect, however, and it was only the increasing shortage of water which led to more positive action in 1818. The new supply was completed in 1823.

Glencorse dam[14] was the joint design of Thomas Telford and James Jardine, an Edinburgh civil engineer. It stands to a height of 77 feet at a suitably narrow point in the river valley and is 540 feet long. A view of the dam is shown in Plate 49. During its working life the dam has not, it seems, experienced any problems or required much maintenance, but its construction was by no means so straightforward. Very rightly, Telford was insistent that the dam be founded on rock, and meeting this requirement involved an unexpected amount of excavation in the gravel and alluvium in the river-bed. Several anxious months passed during 1821 before bed-rock was reached and a solid base located for the dam's puddled clay core wall.

During the dam's construction there was a fear at one stage that a sudden rise in the water level was going to wash away the half-finished dam. In fact the danger receded, but had it materialised, there were plans to protect the works by spreading tarpaulins over the exposed portions of the dam's interior. It is perhaps as well that these measures were never put to the test.

The Glencorse dam, then, has been an eminently sound piece of work, and was only the first of several that continue to function as part of Edinburgh's water-supply system; by the middle of the nineteenth century five more earth dams had been built on much the same pattern.

In England the vast majority of early water-supply works and their associated dams are to be found in the north. Here, after all, were the rapidly growing industrial cities, their locations being not unconnected with the fact that in the Pennines and the Lake District there were supplies of water to serve the very industries around which the towns and cities grew up. These same resources were the ones harnessed for the supply of domestic water.[15]

One of the earliest dams was also one of the biggest. In the 1820s the Great and Little Bolton Waterworks Company was empowered to take water from springs on the moors to the north-west of the town, but only on condition that compensation water was supplied to mill-owners downstream. This stipulation led to the construction of the Belmont dam in 1827. Originally it was 65 feet high and was raised to its present height of 81 feet in 1843; the earth embankment is 1,038 feet long. The Belmont dam, now nearly 150 years old, still functions as part of Bolton's water-supply system, although between 1923 and 1926 extensive repairs were necessary to take care of a slip in the embankment.

In fufilling its role as a supplier of compensation water the Belmont dam led to an interesting development.[16] Local mill operators discovered, quite logically, that water from the dam furnished them with a much more reliable and regular supply than had ever been obtained from the river directly. So, in 1837, they built, entirely for their own benefit, the Entwistle dam which was one of the highest of England's early dams. In fact the intention at first was for a dam 128 feet high, but it was completed only to a height of 108 feet. As a part of Bolton's water-supply system the dam is still in use (Pl. 50) and impounds nearly 3,000 acre-feet of water behind its 360-foot-long earth wall.

By 1840 there were a dozen or more large (of the order of 50–60 feet high) earth dams already built in England as part of water-supply systems. Places such as Bolton and Manchester in Lancashire and Huddersfield and Sheffield in Yorkshire were responsible for most of them. In 1848 Manchester Corporation embarked on the construction of a series of water-supply dams in the Longdendale valley, ten miles east of the city, the first of several much bigger water-supply schemes based on dams which were gradually built up during the second half of the century for cities such as Manchester itself, Liverpool, Birmingham and Sheffield.

By 1850 English earth dam construction was tending towards a standard form. Core walls were almost invariably of puddled clay, underneath which a puddled clay trench was built right down to bed-rock to prevent water percolating under the structure—not a particularly wasteful process but exceedingly dangerous. The earth fill on each side of the core wall was sloped at gradients of about 1 in $2\frac{1}{2}$ to 3 on the water-face side and 1 in 2 on the air-face side. The dam's water faces were covered with masonry and the air faces often sown with grass. In quite a short time, not much more than half a century, British engineers arrived at a form of dam which by and large has proved to be successful. Much the same mode of construction is still used today on occasions, although concrete has displaced puddled clay for the impervious layer. It is true that there have been failures (they will be mentioned in Chapter 9), numerous repairable slips and deformations have occurred, and settlement has often damaged the outlet pipes running through the bases of the embankments; but overall the earth dam in Britain has enjoyed a better safety record than in most other countries, particularly the United States.

One other outcome of the construction of water-supply dams needs to be mentioned; it focused attention on hydrologic questions and also provided some answers. In designing water-supply reservoirs it is obviously important to know how much water is likely to be available from a given catchment area from year to year. Ultimately accurate methods of prediction were developed on the basis of rainfall measurements, but in the early days some rough-and-ready figures were derived from observations of existing reservoirs and the rates at which they filled.

In France only one water-supply dam was built before 1850, but its importance more than compensates for the fact that it is the sole example. The Zola dam[17] was the first arch dam built in France; at the time it was built it was the highest arch dam in the world, a distinction which it retained until 1887 when the Ponte Alto dam was heightened for the last time; and it was the first arch dam which became sufficiently well known to focus international attention on the advantages of this type of structure.

To the east of Aix-en-Provence the River Infernet was first dammed in the seventeenth or eighteenth century by means of a small semicircular dam about 8 feet high which supplied water to a nearby château. It was just upstream from this structure that the Zola dam was built between 1843 and 1854 to supply water to Aix-en-Provence.

The Zola dam (Pl. 53) is built at a narrow point on the Infernet where the valley sides are steep and composed of hard limestone rock. The dam is 36 metres (118 ft) high and curved to a mean radius at the crest of 51 metres (168 ft). But for three horizontal steps, each one a foot wide, the water face is vertical. The crest is 6 metres (20 ft) thick and the air face is steeply sloped in three separate sections: at a gradient of 1 in 5 for the upper 17 metres (56 ft), at 1 in 20 over a central portion 10 metres (33 ft) high and at 1 in 4 over the bottom 9 metres (30 ft). At its base the dam is 13 metres (43 ft) thick. Below this level the structure extends a further 21½ feet down into its foundation trench. For the period, these dimensions indicate a notably thin arch dam.

The Zola dam is made of rubble masonry faced with cut stone blocks, and even today it is in a perfectly watertight condition. It has three outlets. The one at the base was once used when necessary to drain the reservoir but is not now operational; the two others are about half-way up the dam on the left-hand side. The lower one is also a drainage gallery, and the other feeds the Canal Zola which carries water to Aix-en-Provence. The dam is also equipped with an overflow spillway consisting of a channel, 26 feet wide and a few feet deep, cut on a steep gradient into the rock beyond the left-hand end of the dam. The spillway is semicircular in plan view, its intake being just above the dam and its exit just below, about half-way down the ravine. In times of heavy flooding, however, this spillway has insufficient capacity, and a good deal of water finds its way over the dam's crest. Apparently no damage has resulted.

The reservoir's capacity is 800,000 cubic metres (650 acre-feet) of which only about 6 per cent has been taken over by silt during the last 125 years or so. Hence it is not serious that the dam has no scouring gallery, and in any case the recently (1952) completed Barrage de Bimont, a magnificent arch dam 1½ miles upstream, is now the major reservoir in the area.

The Zola dam is unique among dams, and rare among civil engineering structures generally, in that it bears the name of its designer. François Zola, the father of the novelist, Emile Zola, was of Italian and Greek descent but lived for most of his life in France. In 1839 he prepared two designs for a dam on the Infernet, one 30 metres (98 ft) high and the other 48 metres (158 ft) high. Basically they were similar, and there is a suggestion that Zola carried out some calculations in order to select the form of the dams. If this is so, then the Zola dam is unquestionably the first arch dam for which design calculations were made. Sadly François Zola did not live long enough to see his work completed (he died in 1847), and in fact the completed dam is a compromise between the two originally designed. Nevertheless it is altogether a magnificent dam, and a visit is well worth the trouble required to reach the site.

Like France, the United States was at first slow to build water-supply dams, but before 1850 a number had been completed. One of the earliest dams used in the United States for water supply was also a navigation dam. In 1819 the city of Philadelphia in collaboration with the Schuylkill Navigation Company undertook the construction of a dam across the Schuylkill river at Fairmont. The arrangement was that the navigation operators should use the dam as a source of water for locks at one end while the city of Philadelphia installed its water-powered supply pumps at the other end.

The Fairmont dam[18] was built obliquely across the Schuylkill in order to cope with the river's considerable flood flow. Made of wood, earth and stone, the main wall was over 1,200 feet long and raised the river 30 feet. It created a reservoir six miles long and must have been a well-executed piece of work because in February 1822, even before it was commissioned for use, the dam sustained a flood nine feet deep over its long crest. The Fairmont dam appears to have been used for many years, but its eventual fate is not clear.

The most important and biggest of the United States' early water-supply dams was that constructed between 1837 and 1842 across the Croton river, forty miles north of New York. By 1835 New York's demands were greatly in excess of what could be supplied from springs and wells, and it was John B. Jervis who proposed an impounding reservoir on the Croton, a tributary of the Hudson river. From the dam, water was brought to New York along a 42-mile brick aqueduct.

Initially the Croton dam[19] was intended to be a predominantly earth dam with a masonry section at the southern end, the only point on the river-bed at which a rock foundation could be located. A natural rock abutment was available at the southern end of the dam, but at the northern end an artificial abutment had to be constructed.

In 1841 the partially completed earth section was washed away by a flood. It was decided at this point to continue the structure as a predominantly masonry dam with a short earth section at the northern end. Since no rock foundation was available, an artificial one was created. Loose mud and stones were cleared from the river-bed, and a massive wooden foundation, reinforced with masonry and concrete, was constructed. On top of this strange but very substantial foundation the masonry dam was built.

When completed, the Croton dam was 430 feet long and 50 feet high. The full length of the dam's water face was protected with a massive earth bank sloping at a gradient of about 1 in 5 to a base thickness of 275 feet.

Near its top the earth bank was faced with stone as protection against overflow, this being allowed to discharge clean over the central 180-foot length of the masonry portion of the dam. The structure's profile was specially shaped to carry the overflow, and a downstream wooden apron was built to prevent the structure from being undermined.

Further protection against the effects of overflow was derived from another dam. A short distance downstream a small timber dam was built which flooded the space between itself and the main dam. This helped to break the force of water pouring over the Croton dam and also served to keep the latter's wooden foundations saturated—a shrewd move. The Croton dam was fitted with a single outlet tunnel which was used to supply the aqueduct to New York and also for draining the reservoir.

The Croton dam served New York for the remainder of the nineteenth century, but then it suffered the supreme indignity. Between 1892 and 1906 a new and much bigger structure called the New Croton dam was built $3\frac{1}{4}$ miles downstream. The original structure became known as the Old Croton dam and was then 'buried' in the new reservoir.

It was pointed out earlier that one of England's earliest large dams, the Entwistle dam, was built by mill-owners as a source of supply to their water-wheels. It is worth reflecting at this point on the question of water-power dams generally between 1800 and 1850.

Existing accounts of the history of engineering in Britain during this period give a great deal of space to the use of the steam engine as a prime mover. The work of Newcomen, Watt and Trevithick was, of course, of epoch-making importance and one of the most significant technical accomplishments of the eighteenth and early nineteenth centuries. But the notion that steam engines alone powered the machines of early industrial Britain and immediately replaced all utilisation of water power is certainly not true. It would be intriguing to know what percentage of stationary industrial horse-power was being generated by steam engines in, say, 1800; even in Britain it was perhaps not so much as is generally imagined, and in Europe and North America it was very much less. P. N. Wilson has prepared a revealing table of statistics[20] which shows that in British cotton mills in 1834 one-third of the power generated came from water-wheels.

There is no doubt that in Britain in the first half of the nineteenth century water-power resources continued to be developed and a number of dams were built for the purpose.

Until Lake Thirlmere's completion in 1894 the largest reservoir in Britain was Loch Thom, named after its builder, Robert Thom, and designed partly to power the mills of Greenock. (Ironically it was at Greenock that James Watt was born.) Loch Thom was created in 1827 by the construction of an earth dam, of typical form, 1,400 feet long and

66 feet high. Four much smaller embankments were used to plug gaps in the sides of the reservoir space. Water was taken from the reservoir to Greenock along a six-mile aqueduct with a fall of 512 feet. It is claimed that a supply of 20 cusecs was provided, and the reservoir by itself could sustain Greenock's mills for six months in a dry spell.

An equally impressive scheme was prepared but never completed for the River Kent and its tributaries the Mint and Sprint, in Westmorland. The engineer was J. F. Bateman who in 1836, along with Sir William Fairbairn, had worked on the problem of developing water power from the River Bann in Northern Ireland and had built the Lough Island Reavy reservoir in the mountains of Mourne. In Westmorland, Bateman proposed six reservoirs with a combined capacity of 7,000 acre-feet. In fact only one reservoir was built, the one at Kentmere Head, and then around 1850 the scheme was given up. The River Irwell in 1835 had some 300 mills already at work, and here too plans for water-power dams were never realised. Had they been, an estimated 6,600 h.p. would have been available.

While the above projects in north-western England fell through, in the Pennines others forged ahead. Earlier in this chapter mention was made of a 'contour dam' near Nenthead called Perry's dam, which was built for the benefit of local lead mines. It was by no means an isolated example, and together with dams of the conventional type—those built across streams rather than on the faces of hillsides—lead-mining interests were responsible for the building of some dozens of water-power dams in the north of England.

Arthur Raistrick called the period 1780–1880 the 'age of mechanisation' in English lead mining, and of this hundred-year period the middle fifty years seem to cover the construction of most of the dams.[21] A particularly fine set were built on Blea Beck for the mines on Grassington Moor, and were part of an elaborate system of water channels and water-wheels which once powered grinding mills and winding engines.

Of the three dams on Blea Beck, the lowest is the largest, an earth bank 25 feet high and 450 feet long. The water face has a rough masonry lining and there are two outlets, a high-level sluice at one end feeding a long surface aqueduct leading to the mines, and a low-level sluice for drawing down the reservoir. The two other dams are similar but smaller, and neither is now in use. In fact at their left-hand ends both dams are breached right down to foundation level. As a result it can be seen that both are very simple pieces of construction. There is no core wall, no cut-off trench, nothing but a plain earthen bank with a rough masonry facing. The same type of construction can be seen in the Swinehopehead dam on Swinehope Burn near Allenheads, which has also been breached, presumably to make it safe. This structure is 500 feet long, 15 feet high and some 50 feet thick at the base.

Near Stanhope is another group of mine dams of which one, Burnhead dam, is particularly large and still in use. Along the crest the dam is 10 feet wide and 500 feet long. At the base it is about 100 feet thick and stands nearly 40 feet high at the centre. The old low-level outlet is now hopelessly choked with sediment brought down off the moors; but there is a large crest-level spillway—a feature common to nearly all Pennine mine dams.

These dams cannot be considered as anything but a very minor part of the overall picture. It should be remembered, however, that they were built by small groups of miners anxious to increase their yield of lead and thereby their profits. Mechanisation had to be based on water power since steam engines were an impossible proposition under the prevailing conditions. Today these old Pennine mine dams are in their way an impressive and lonely memorial to a once thriving industry.

Throughout the nineteenth century the water-power resources of the United States were steadily developed, principally in New England and California. The dams and water-mills built in the New England states in the early part of the century were a logical development of the pioneer schemes mentioned at the end of Chapter 6.

A particularly well-known development was at Lowell on the Merrimack river in Massachusetts. One of the earliest water-power schemes here utilised an already existing dam which belonged to the proprietors of the navigation on the river. In 1822 a new dam was built, one of the biggest of timber to have been constructed up to that date. It was used to drive textile mills and is said to have been capable of delivering up to 12,000 h.p. Lowell rapidly became a major textile centre depending for its power on more dams and even more water-wheels. Other Massachusetts towns were drawn into the textile-manufacturing business, and also paper-making, so that by the middle of the century there were dozens of dams and mills scattered about New England. The area was, in short, a world centre of water-powered industry.

Lowell's most important contribution to engineering was not in dams, however. The chief engineer of one of Lowell's groups of manufacturers was James Bicheno Francis, and it was at Lowell in 1847 that Francis carried out his pioneer researches into the performance of the outward-flow turbine and subsequently an inward-flow type.[22] As we shall discuss in Chapter 9, this work marked the opening of a new era in water-power utilisation and was partially responsible for important new developments in dams towards 1900.

In order to power their paper mills, the Holyoke Water Power Company in 1849 built one of the largest of all rubble and timber dams.[23] Initially the Connecticut river was dammed at Holyoke, thirty miles upstream from Hartford, by means of a rough wooden dam which was intended to provide temporary water power and also serve as protection for the permanent dam while it was being built. In November 1848, when it was filled for the first time, this unsubstantial structure was rolled over and floated down the river on the crest of the flood it had released.

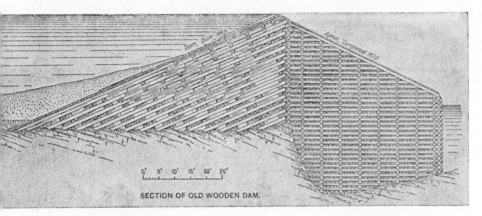
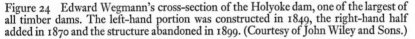

SECTION OF OLD WOODEN DAM.

Figure 24   Edward Wegmann's cross-section of the Holyoke dam, one of the largest of all timber dams. The left-hand portion was constructed in 1849, the right-hand half added in 1870 and the structure abandoned in 1899. (Courtesy of John Wiley and Sons.)

The more permanent structure built in 1849 was 1,017 feet long and 35 feet high. Its cross-section is shown in Figure 24 above. It can be seen that the dam was built in two distinct pieces, one on each side of a natural ledge across the river-bed. None of the timber beams used in the structure were less than twelve inches square and those at the base were larger. In all the dam contained 4 million feet (board measure) of timber, a prodigious quantity of wood which was cut and placed in just six months.

The base of the dam was pinned to the rock bed of the river with 3,000 iron bolts. As the dam was raised the spaces between the beams were filled with rubble and stones, and finally the crest was faced with iron sheets. A layer of earth was dumped at the foot of the sloping water face.

The Holyoke dam had a chequered career, mainly because it had to discharge so much water over its crest. Between 1868 and 1870 the portion of the dam below the rock sill was constructed to prevent further undermining of the foundations. In fact this modification merely shifted the problem to the downstream edge of the enlarged dam. Gradually the dam became less and less reliable, severe leaks being added to the effects of erosion, and it was finally replaced in 1899, fifty years after its construction.

During the first half of the nineteenth century the beginnings of a new trend in dam-building can be detected. It concerns the reappearance of dams for irrigation in North Africa and Asia, and in the case of Egypt their construction for the first time. It would be wrong to give the impression that in the preceding centuries the art of dam-construction had been entirely dead in these regions, but certainly contact with European influence and European engineering stimulated new activity.[24]

In the early decades of the nineteenth century the basin irrigation system on the Nile was proving increasingly inadequate as Egypt's population grew. The then viceroy, Mehemet Ali Pasha, resolved to develop Egypt's economy and selected cotton as the crop which could best be exported in exchange for materials, food and money. A serious problem had to be faced, however: to grow cotton in Egypt requires irrigation in the spring and summer, precisely the seasons when the Nile is low. It was decided therefore to dam the river at the head of its delta and thereby irrigate the land between this point and the sea. Even before 1800 no less a figure than Napoleon had appreciated that such a scheme would one day be necessary if the Nile delta was to be properly developed, and the river's two delta branches fully controlled.

With more enthusiasm than ability, Mehemet Ali's engineers began work on a dam[25] in 1833. At one stage it was intended to obtain masonry for the structure by dismantling the Pyramids, but fortunately this absurd and vandalous scheme was soon abandoned on account of the expense it would have entailed. Then in 1835 a plague wiped out the bulk of the labour force and the project was given up.

In 1842 a new scheme was prepared by two French engineers, which called for the construction of two dams very close together, one on each of the Nile's delta branches. The Damietta barrage was built first, and its construction apparently went through without difficulty. It is 535 metres (1,760 ft) long. The dam on the Rosetta branch was begun in 1847 and finally completed fourteen years later to a length of 465 metres (1,530 ft). Both dams were equipped with continuous series of sluices with which to control the flow of the Nile and to ensure that silt, of which the Nile carries vast amounts, did not collect above the dams. Navigation was provided for by sets of locks in both dams.

The Rosetta and Damietta dams were the first ever built on the Nile, and while the projects were entirely sound in principle, they were shoddily executed. The Rosetta dam in particular leaked badly through its foundations, and in 1867 a large section of it slipped perceptibly downstream. Numerous inspections of the dams by British engineers led ultimately (1883) to the appointment of C. C. Scott-Moncrieff as chief engineer of Egypt's irrigation works, and the dams were soon made safe and reliable.

Scott-Moncrieff and others—such as Sir William Willcocks, later to plan the rehabilitation of Iraq's irrigation—who worked on dams in Egypt had learned much of their irrigation engineering in India. It was to this country that British civil engineering skill was first exported on a large scale, although the problems which had to be faced there were not like those in Britain. It is worth remarking here also that all the really big dams built by British engineers have been built abroad, and in the nineteenth century it was India which claimed the pick of the crop.

One of the first examples of British influence on the dam-building story in India concerns the Coleroon river in Tanjore.[26] As far back as the second century A.D. the Chola kings had constructed a dam across the River Cauvery near its mouth. Made of masonry laid in clay, 1,080 feet long, 40–60 feet thick and from 15 to 18 feet high, it is among the most remarkable of India's very many old dams and tanks still in use.

By 1830 the ancient works—dam and canals—were heavily silted, and most of the Cauvery's flow was being diverted into the neighbouring Coleroon, one of the Cauvery's delta outlets. Repairs and modifications to the ancient dam were not enough, and in 1836 Arthur Cotton built a new dam on the Coleroon. It is interesting that Cotton followed closely the style of the ancient dam, even daring to build the new structure on soft alluvial foundations just like its precedessor. He was not entirely successful. The dam partially failed in its first year, was repaired, and remained in a risky condition until it was rebuilt at the end of the century. The Coleroon dam is of historical interest, however, because its un-satisfactory performance drew attention for the first time to the phenom-enon of piping, the process by which water percolating under a dam has sufficient pressure and velocity to erode the foundations, often to the extent that the dam is undermined. Similar failures of other Indian river dams eventually led to the first experimental investigations of seepage under masonry dams, and the work of Beresford and Clibborn in the 1890s was able to predict the failure in 1898 of the Narora weir on the River Ganges.

The energy and perseverance of Arthur Cotton was the mainspring of the earliest British dam-building exploits in India; around 1850 he also built dams on the rivers Godavari, Kistna and Mahanadi. Later on, in the last third of the nineteenth century, storage dams appeared in India, and these mark a logical and necessary (and also expensive) step beyond the earlier diversion schemes.

This chapter can usefully be concluded with a mention of an Indian dam which is not well known but was, at the time it was built, a most original and imaginative piece of work.

The Meer Allum dam (Pl. 54) was built to supply water to Hyderabad.[27] Little seems to be known about its origins. The date of its construction is variously placed at *c.* 1790–1810, and different writers have attributed

it to British, French or Indian engineers. Whoever was responsible, the
dam reflects credit on their design and workmanship.

The Meer Allum dam has a maximum height of 40 feet and is some
2,500 feet long in the form of a long curve. This wall is divided into 21

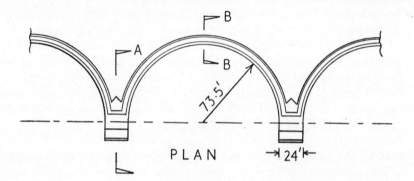

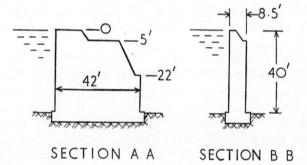

MEER   ALLUM   DAM

Figure 25   Some details of the Meer Allum dam near Hyderabad.

semicircular arches whose spans range from 70 feet at the ends to 147
feet near the centre. Each arch has a constant thickness of 8½ feet but for
a stepped portion at the crest. The arches are supported by a series of
buttresses; each one is 42 feet long and 24 feet thick. The tops of the
buttresses are level with the crests of the arches where the two meet, but
are stepped at their downstream ends.

The Meer Allum dam is the earliest-known example of a true buttress
dam of the multiple-arch type. The form is much more fully developed

than in any of the dams of Don Pedro Bernardo Villarreal de Berriz in Spain, and the structure is very much larger. Indeed its size is remarkable, bearing in mind that the idea had never been tried before.

The reservoir impounded has a capacity of nearly 8,000 acre-feet of water, and overflow is discharged partially through a spillway at one end while the rest pours over the crest. In more than a century and a half this has not damaged the masonry and mortar of which the dam is built, nor has it undermined the foundations.

The first half of the nineteenth century is essentially a period when dams were built in markedly increased numbers but with little improvement in design and still no proper understanding of their structural behaviour.

The masonry dam, whether arched or gravity, remained a structure built on the basis of empiricism and experience, although this situation was to change, dramatically in the case of gravity dams, in the next half-century. The earth dam, widely used for canal and water-supply works, was reduced to a standard and generally reliable form, especially in Britain, but any real understanding of its behaviour or appreciation of the special problems associated with retaining high heads of water behind earth banks was lacking. The timber and rock dam had its heyday in the United States but became increasingly less common after 1850; it proved ultimately to be too vulnerable. Moreover, water-power installations based on diversion dams were eventually replaced by hydro-electric schemes.

By 1850 canal dams had virtually ceased to be built. Railways provided the basis of most countries' transport systems, and subsequent canal schemes, such as Panama, Suez and Kiel, were of a different type and designed for a different purpose.

Reservoir dams for water-supply schemes have their origins in the early nineteenth century and, ever since, have been built the world over when the available resources require them.

The entry of Great Britain and the United States into the ranks of the major dam-building countries dates from the period 1800–50.

The Second Half of the Nineteenth Century

T HE DEVELOPMENT WHICH dominated the evolution of dam-building more than any other in this period was the formulation of methods of stress-analysing masonry dams and the application of these methods to the design of new structures. Attention was focused principally on gravity dams, but some attempt was also made to develop the means to work out stresses in arch dams. The consequences of this theoretical work on dams were profound, and what for centuries had been essentially an art was given a scientific dimension for the first time. The immediate result was the appearance of dam profiles of quite different shape. Indeed, there is probably nowhere in civil engineering a more striking example of a structure's basic shape being so radically altered by the application of a rational theory; and the change occurred in a remark-ably short time. By the end of the century the analysis and design of masonry dams, while far from perfected and still the subject of dis-agreement and even controversy, was close to being placed on a sound footing and one from which twentieth-century developments and refine-ments stemmed.

Although the first important work on the structural analysis of dams was done after 1850, there is what might be called a pre-history of dam design which should be reviewed briefly.

An initial calculation which needs to be made when attempting any theoretical study of dams is the one that determines the position and total intensity of the applied water load. It is a simple problem in hydro-statics. The solution was first worked out by Simon Stevin and published in his *De Beghinselen des Waterwichts* of 1586.[1] Stevin showed that the water pressure acting against a vertical surface varies uniformly from zero at the top to a maximum at the deepest point, i.e. a triangular pressure diagram. From this he calculated the magnitude of the applied force and correctly showed that its point of application is at two-thirds the depth of the water.

Stevin also published a few general observations concerning the foundations of dams.[2] He was particularly interested in small masonry dams built on pervious foundations, a situation which he realised could lead to a dam being destroyed by the water percolating underneath it. His solution, which he described with drawings, entailed the construction of deep wooden foundations, and for these Stevin recommended the use of dovetailed piles for the outer layer. He was satisfied that such an arrangement was essential to the security of any dam built on a material such as sand. Stevin appears to have been the earliest writer on the theory and practice of dam-building, although it is not clear to what extent his ideas were applied.

Bernard Forest de Belidor was the first person to publish calculations relevant to the determination of the stability of dam walls. In the Second Part of Volume 1 of his *Architecture Hydraulique*, published in 1750, he discussed[3] the resistance to overturning of a masonry wall of rectangular cross-section when loaded with water on one side.

Belidor considers a wall whose height is b and whose thickness is y at all levels. He takes a depth of water a, less than b, and calculates how thick the wall must be in order to ensure that the water load will not overturn the wall about its air-face base corner, often referred to as the toe of a dam. By equating the overturning moment due to the water pressure with the restoring moment developed by the weight of the wall, Belidor shows that the minimum required thickness is given by $y = \sqrt{\dfrac{a^3}{5b}}$; he assumes that masonry is $1\frac{2}{3}$ times as heavy as water—a low figure but erring on the side of safety. It is interesting, too, that Belidor makes a further concession to safety by suggesting the use of a safety factor of 1·5. He applies it, however, not to y but to the density of the wall. Thus he reduces the effective density of masonry to $\frac{10}{9}$ that of water and hence finds y to be $\sqrt{\dfrac{3a^3}{10b}}$. In effect he has increased y by a factor of $\sqrt{1·5}$, the weight of the wall in the same proportion and hence their product by a factor of 1·5, the intended safety margin.

It is not clear whether Belidor's method of calculating the thickness of a water-retaining wall was subsequently used by dam-builders, but since the *Architecture Hydraulique* was so well known in Europe it may well have been.

In the eighteenth century, while Belidor dealt with an aspect of dam design based on rational mechanics, a number of other writers propounded 'theories' of a more qualitative type. They amount essentially to sets of general instructions on how to construct dams, usually small diversion works, and the dimensions necessary to ensure an adequate structure. The notions put forward were based on a mixture of empiricism, intuition and experience.

One of the writers in question was Don Pedro Bernardo Villarreal de Berriz whose dams were discussed in Chapter 5. Don Pedro's instructions[4] to dam-builders cover two types of structure: straight gravity dams and the particular type of multiple-arch dam which he himself developed. In the former case Don Pedro recommended a masonry structure with a vertical water face and an air face sloping at 45 degrees. For his multiple-arch dams Don Pedro prescribed simple geometrical rules to fix the dimensions. The width of a buttress should be a quarter of the chord length of each arch; the thickness of the dam at a buttress should be twice the dam's height; the water face should be straight and vertical and the crest flat. The radius of each arch is recommended to be equal to the arch's chord length.

Don Pedro gives meticulous instructions on how to construct dams and is most particular that well-prepared materials be employed in a proper fashion. He emphasises the need for carefully mortared joints to ensure a watertight and permanent structure. In the case of straight gravity dams Don Pedro suggests that iron clamps be used to fix the masonry blocks, and that the sloping air face be protected with wood.

A similar empirical text on dams was published by Charles Bossut. In his *Recherches sur la construction la plus avantageuse des digues* of 1764, Bossut considers the use of dams to create small reservoirs for water-power installations. He is very much in favour of structures made of masonry and mortar, and whenever hard rock cannot be located he recommends the use of piled foundations. The same arrangement is prescribed for earth dams under certain conditions, and the need for a masonry facing on the upstream side is emphasised.

Charles Bossut's book pays a good deal of attention to hydraulic problems and warns of the necessity to provide adequate spillways in dams of any type. Moreover he appears to be the earliest writer to deal with spillway profiles and the trajectories followed by streams of water pouring over the crests of dams.[5] The latter Bossut considers in conjunction with downstream aprons designed to prevent undermining of the foundations. Another eighteenth-century work *Traité sur la science et l'exploitation des mines*, written in 1778 by Christofe Delius, also reflects this same concern with dam hydraulics.

Before 1850, then, a number of texts dealing wholly or partly with the design and construction of dams had been written, but only Belidor's made recourse to a rational mathematical approach.

In 1853 M. de Sazilly, a French engineer, proposed a method of gravity dam analysis based on the idea, outlined in Chapter 2, that a gravity dam can, in the first instance, be visualised as a series of separate vertical slices each of unit width (say 1 foot or 1 metre). Starting from this concept, de Sazilly's analysis[6] then assumes that any such vertical

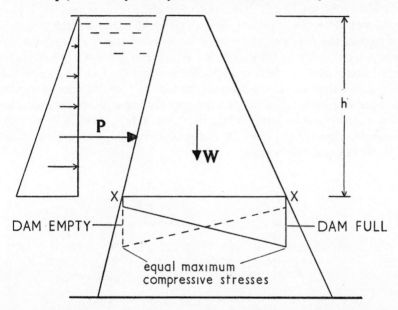

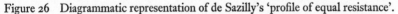

Figure 26   Diagrammatic representation of de Sazilly's 'profile of equal resistance'.

slice behaves like all the others and can be treated as a vertical cantilever, fixed at the base and loaded by the water pressure along its upstream side. The problem is shown in Figure 26, the profile illustrated being reduced to a simple form. It will be seen that in essence de Sazilly treats the problem as a particular case of beam-bending, a topic which had been generally clarified in 1826 in Navier's *Résumé des Leçons de Mécanique*.

M. de Sazilly based his theory of gravity dams on two propositions. At any horizontal section, such as X-X, the weight and shape of that part of the dam above X-X was to be such that (*a*) it could not slide horizontally, and (*b*) the maximum vertical compressive stress on X-X should not exceed a predetermined safe limit. De Sazilly points out that in determining the maximum compressive stress two extreme cases must be considered. When the reservoir is empty, the maximum compressive stress will occur at the water face, and the diagram of vertical stress distribution on X-X will be as shown. It is due entirely to *W*, the weight of masonry above X-X. When the reservoir is full, however, the horizontal force P, due to the water pressure, will modify the stress diagram because a bending moment equal to $\dfrac{Ph}{3}$ will be introduced at the section X-X. This will cause the maximum compressive stress to develop at the air face according to the second trapezoidal stress diagram shown.

It is important to recognise here that any theory of dams faces two distinct problems. One is to analyse the stability of and stress levels in an existing profile; the other is to design a safe profile from scratch, knowing, of course, the depth of water to be impounded and the maximum safe stresses which can be allowed. It was in order to develop a procedure for designing a dam profile that de Sazilly imposed the following condition. He stated that at each horizontal section X-X the maximum vertical compressive stress when the dam is empty shall be equal to the maximum vertical compressive stress when the dam is full. This in fact is the condition depicted in Figure 26, and produces a design having a so-called 'profile of equal resistance'.

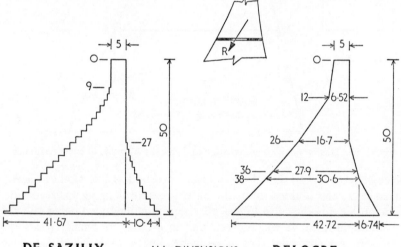

DE SAZILLY   ALL DIMENSIONS   DELOCRE
                  IN METRES

Figure 27   A comparison of the profiles of de Sazilly and Delocre. Also shown is the 'middle third' condition.

De Sazilly attempted to find formulae which would enable him to calculate the profile of equal resistance exactly. But although he obtained the correct differential equations he was unable to integrate them. So he dodged this problem by using a profile which was stepped on both faces and therefore yielded elements of finite size which could be calculated one at a time (Fig. 27).

The profile of equal resistance was actually the whole basis of de Sazilly's approach. His other postulate, namely that sliding was to be avoided, was largely ignored, since he knew of no dam which had experienced trouble of this sort. It is interesting that de Sazilly makes no formal mention of two other conditions which need to be met. One is that no tensile stresses should be allowed to occur, and the second is that a dam must be able to resist overturning about the toe of any section as envisaged by Belidor. In fact the profile of equal resistance automatically covers these requirements, but it is not clear whether de Sazilly realised this. Nevertheless he must be credited with the first research into the stress analysis of dams.

De Sazilly's theoretical work on masonry dams is indicative of two things. Firstly, France was the country in which structural analysis had up to that time experienced by far the greatest degree of its development. Secondly, there was in France a decided lack of faith in earth construction for large dams compared with masonry, a situation perhaps connected with French experiences with canal dams in the early part of the nineteenth century. Earth dams had had on the whole a less distinguished record than masonry dams.

In 1858 the French Government decided to construct a large reservoir on the River Furens to provide flood protection for Saint-Étienne.[7] For effective control of the river a dam of unprecedented height, 50 metres (164 ft), was required, and inevitably the engineers in charge of the project elected to build a masonry structure. Quite apart from French distrust of earth dams, their decision was possibly influenced by the already existing masonry dams in Spain, particularly that at Alicante, which at the time was the only dam in the world comparable in height to what was intended for Saint-Étienne. Moreover, the two sites were themselves of similar shape.

The engineers of the Furens dam (Pl. 51), Messieurs Graeff and Delocre, aware that existing masonry dams were unnecessarily massive and by nineteenth-century standards would prove expensive to build in terms of both time and money, adopted de Sazilly's ideas as the basis for their designs. The chief designer was Delocre.

M. Delocre added nothing new to de Sazilly's two canons of design, although he did away with a profile having stepped faces. He argued that this not only involved a waste of material but required in addition an expensive type of masonry with which to construct the steps. Delocre proposed, therefore, a profile with polygonal faces, the straight sections changing direction at the quarter-points of the dam's height. In this way Delocre saved 33·5 cubic metres of masonry per metre length of the dam. Figure 27 shows a comparison of the de Sazilly and Delocre profiles.

Delocre's paper[8] on the design of dams in general and the Furens dam in particular is of considerable importance, and it contains an interesting and basic oversight. To determine the maximum vertical compressive stresses in a dam Delocre locates at any horizontal section (Fig. 27) the point of application of the resultant pressure R (it is due to W alone when the dam is empty and to P and W combined when the dam is full) and he notes in passing that, if R is outside the middle third of the section, then tension will occur at one face or the other (in practice it occurs at the water face with the dam full). But apart from 'neglecting the force of cohesion in the mortar, which is unfavourable to resistance', Delocre is unimpressed by the possible presence of vertical tensile stresses and makes no statement that they should be avoided.

De Sazilly makes no attempt in his paper of 1853 to specify the maximum compressive stress which should be allowed in a gravity dam. Delocre, however, being involved in the design of a structure which was to be built, was obliged to decide on a maximum stress value on which to base his calculations. This led to an interesting line of reasoning. M. Graeff in his paper[9] on the Furens dam confirms that French engineers were aware of existing masonry dams in Spain and, of course, those in France as well. In order to determine the stresses which these old dams had safely sustained, for 200 years or more in some cases, they were all analysed by the de Sazilly method. Six old Spanish dams—Puentes, Valdeinfierno, Nijar, Almansa, Elche and Alicante—were found to be stressed to maximum values ranging from 6·5 kg/cm.² (92 lb/in.²) in the case of Valdeinfierno to 14 kg/cm.² (197 lb/in.²) in the case of Almansa. For two French dams, Bosmeleac and Grosbois, the maximum vertical compressive stresses were found to be 6·09 kg/cm.² (86½ lb/in.²) and 10·4 kg/cm.² (148 lb/in.²) respectively. These figures settled the upper and lower design limits for Graeff and Delocre. A maximum stress of 6 kg/cm.² (85 lb/in.²) was regarded as amply conservative and 14 kg/cm.² (197 lb/in.²) as the limit beyond which it was unsafe to go.

There are a number of curious and revealing aspects in the above procedure. In the first place it is surprising that design limits were taken from analyses of existing structures and not from tests on materials, the very same materials of which it was intended to build the Furens dam. Secondly the lowest limit of stress, 6 kg/cm.², is in fact very low indeed for masonry, and even the upper limit of 14 kg/cm.² can be borne comfortably. It remains odd that the designers should have determined and used such low values for dam design and yet did not comment on the fact. A third point emerges clearly, namely that the engineers responsible did not realise, or did not admit, the limitations of the elementary theory they had developed. Their assumption that the stress distributions in a gravity dam are linear is not valid even for a very slender structure, while to apply this notion to such massive structures as Valdeinfierno and

Alicante was utterly inappropriate. In fact later and more accurate methods of stress analysis have shown that the earlier and so-called 'classical methods' give acceptable stress values at most points in the usual gravity dam profile. This, however, had to be demonstrated; Delocre, Graeff and others in the nineteenth century did not even appreciate that such a verification was either necessary or possible.

M. Delocre's design for the Furens dam was based on the lower stress limit of 6 kg/cm.^{2}; he prepared several profiles of as near equal resistance as a reasonable amount of calculation would allow and upon these the construction of the dam was based. Since the dam has a profile of equal resistance, the no-tension and no-overturning conditions are automatically covered. It is worth remarking, however, that M. Graeff, in order to emphasise the economy to be gained by using the higher stress limit of 14 kg/cm.^{2}, shows[10] a profile in which this stress occurs at the air face when the reservoir is full but *not* at the water face when it is empty, i.e. a profile of *unequal* resistance. This results in a line of pressure for a full reservoir which is well outside the middle third at virtually every level, producing therefore considerable tension at the water face. This would seem to confirm the general unawareness that criteria other than those based on maximum compressive stress concepts are required.

Apart from the factors already mentioned, the Furens dam appears to owe one other debt to its Spanish predecessors. As M. Graeff himself observed, 'In France, we do not know of any dams constructed to a curved form: in Spain they are nearly all constructed in this way.'[11]

Why he was unfamiliar with the Zola dam is a mystery, but nevertheless the arched form, so common in Spain as we have seen, was incorporated into the Furens dam. At the crest it is curved to a radius of 252 metres (830 ft). Standing to a height of 50 metres (164 ft), 5 metres (16½ ft) thick at the crest and 49 metres (161 ft) thick at the base (Fig. 28), the Furens dam still serves Saint-Étienne and has recently celebrated its centenary (its construction was begun in 1861 and completed five years later). Still in perfect condition, it is an impressive and singularly appropriate memorial to its builders and the engineers who made the first attempts to design dams.

The design, construction and successful operation of the Furens dam revolutionised masonry dam-building. There was no going back to the old empiricism, the greater economy of the new shapes making certain of that. Engineers quickly assimilated the ideas of de Sazilly and Delocre into their work, but the design method was not applied rigorously in every case. Occasionally this led to serious consequences, as we shall see.

Three big curved masonry gravity dams were built in France soon after the one across the Furens, and as replicas of it. The Ternay dam (Pl. 52) was built to a height of 34 metres (112 ft) between 1865 and 1868 to provide flood-protection and a water supply for Annonay. Between 1867 and 1870 the Ban dam, 46·3 metres (152 ft) high, was constructed to supply water to Saint-Chamond, and in the 1870s the Pas du Riot dam, 34·5 metres (113 ft) high, was added to the water-supply system of Saint-Étienne, two miles upstream from the Furens dam. It is significant that

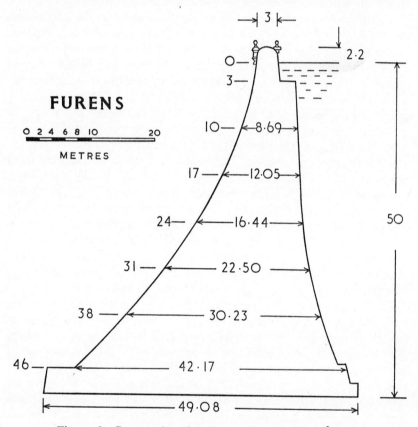

Figure 28   Cross-section of the Furens dam near Saint-Étienne.

for the Ternay dam the compressive stress limit was raised to 7 kg/cm.^{2} (100 lb/in.^{2}), and in the Ban dam it was increased a stage further to 8 kg/cm.^{2} (114 lb/in.^{2}).

In Chapter 8 it was pointed out that British engineers in the nineteenth century made important contributions to dam-building in India, initially with barrages and later on with masonry storage dams for irrigation and water supply. The first of the latter appears to have been the Poona dam, begun in 1868 and built to a height of 98 feet. Others, such as the dams of Tansa, Bhatgur and Betwa, were not completed until the 1890s.

As early as 1808 there was a plan for an irrigation scheme on the River Periyar in Madras; it was finally realised in 1898. The project included the construction of a high masonry dam, and in 1870 the opinion of Professor W. J. M. Rankine was sought as to the form of profile which ought to be used. In a paper[12] published in 1872 Rankine declared himself to be in agreement with the ideas of de Sazilly, Graeff and Delocre, but he saw immediately what was missing from their theories. He points out first of all that the limiting compressive stress concept which governed the French method was based on the determination of the maximum *vertical* compressive stress on any horizontal section. Rankine appreciated that what in fact was required was the determination of the maximum compressive stress *regardless of direction*, this stress rarely being a vertical one but at the same time a higher one than that which was currently regarded as critical. Rankine's proposal to deal with this point was basically empirical, and he did not indicate how to calculate what is in fact the maximum principal stress at the face of a dam.

Rankine's second observation was equally perceptive. He stated that it was essential to avoid tensile stresses in a dam, and pointed out that this would be achieved if the resultant line of pressure was confined to the middle third of every horizontal section for all water levels.

Other less fundamental ideas of Rankine's can usefully be mentioned. He did not propose that the maximum compressive stresses in dams should be significantly increased beyond what was current practice. He was entirely in favour of building gravity dams with curvature because he believed that this would prevent horizontal tension on the air face. He proposed, also, a method of his own whereby the profile of a dam might be designed. This was based on logarithmic curves for both faces, a procedure which was straightforward for the specific conditions dealt with by Rankine but cumbersome to adapt generally.

Rankine's contributions to the theory of dams—one of the last and also one of the best pieces of research he carried out—were of first importance and put the matter on a much sounder footing. Three basic rules now governed the design of gravity dams: (*a*) predetermined maximum compressive stresses should not be exceeded in any direction at any point; (*b*) no tensile stresses should occur, the 'middle third rule' being observed to deal with this requirement; and (*c*) a dam should be heavy enough to prevent sliding. The no-overturning condition was automatically met by (*b*).

Not until 1895, when the problem of uplift was tackled for the first time, was there a further significant addition to these basic concepts of dam design, although before that date a number of analysts—such as Bouvier, Guillemain and Hétier—advanced variations on the existing theme.[13] In the main these were intended either to produce more accurate estimates of maximum stresses, usually by considering oblique rather than horizontal sections, or else were aimed at providing more manageable methods for designing profiles. In fact, however, none of the former were any better as a means of stress analysis, and a design method such as Wegmann's,[14] while more workable than most, continued to involve designers in lengthy calculations and graphical constructions.

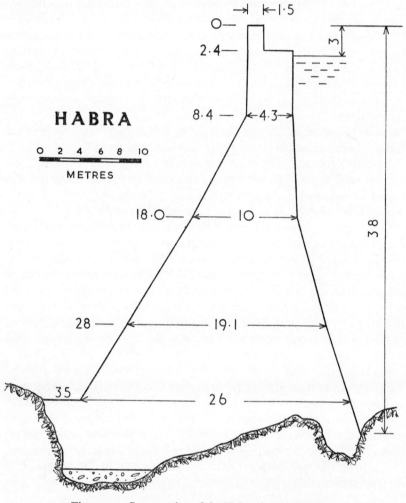

HABRA

0 2 4 6 8 10

METRES

Figure 29   Cross-section of the Habra dam in Algeria.

While the Spanish tradition of masonry dam-building had its effects in France, so it did as well in a country under French rule. It was the Spanish example which encouraged the French government to build a number of large irrigation reservoirs in Algeria and on basically the

Spanish pattern. The earliest, the reservoirs of Habra and Tlelat, were built between 1865 and 1870, immediately after the publication of Maurice Aymard's book which, as noted in Chapter 5, was directed specifically to the attention of the Algerian governor-general. Three more large dams were completed before the end of the century: Djidionia (1875), Gran Cheurfas (1884) and Hamiz (1885).

Apparently all these dams were designed as profiles of equal resistance, but only in the case of the Habra dam was the procedure properly applied. Subsequent to their construction, the others were all shown to be subject to tension on their water faces, and this was presumably the cause of the Gran Cheurfas structure suffering a partial collapse in 1885.

The more famous failure, however, was that of the Habra dam in 1881, when 328 feet of the structure—it was 1,066 feet long and 117 feet high —failed completely at a time when the inadequate spillway allowed the water level to rise nearly thirteen feet above the intended maximum. From the outset the Habra dam[15] had been suspiciously pervious, and the additional pressure caused by the rise in water level is generally believed to have produced uplift on the water face. This effect was undoubtedly encouraged by the development of tension cracks due to the increased load on the dam, and ultimately the reduced thickness of the structure at the level of the cracks produced increased compressive stresses at the air face and dangerously high shear stresses across horizontal sections. Indeed, M. Clavenad in his report of 1887 suggests that the Habra disaster was basically a shear failure, and this persuaded him to adopt shear as the criterion of a design method which he developed. In fact the disaster showed clearly that the mechanism of a dam's failure is very complex, that a whole series of effects occur in quick succession, and that the one which starts the chain is not necessarily the one associated with the final moments of collapse.

The Habra dam's collapse was widely reported but did not create the same degree of concern and even panic which followed the famous failure in 1895 of the Bouzey dam near Épinal in France.[16] Built between 1878 and 1881 to supply the Canal de l'Est, this structure was 1,700 feet long and 49 feet high. As with the Habra dam, its inferior construction led to severe leakage, cracks developed, and in 1884 a large part of the Bouzey dam slipped downstream. Extensive repairs were carried out, but in 1895 nearly 600 feet of the wall was destroyed to a depth of 33 feet below the crest. Much property was flooded and ruined and 150 people were killed.

The Bouzey dam was a straight structure which was claimed to have been designed as a profile of equal resistance (Fig. 30). In fact it was nothing of the sort, not even after its reconstruction and strengthening in the years 1884–9. Throughout its life it was subject to tension on the water face, and especially at the level at which the failure occurred. Subsequent enquiries revealed other defects. The structure's foundations were imperfectly constructed on sandstone, itself permeable and fissured, and the dam wall had not been carried right down to bed-rock. These factors were a primary cause of the dam's early tendency to leak and slide. It was also noted that the Bouzey dam, like the Habra structure, was, for its height, very long. It was astutely observed that this rendered both dams particularly susceptible to thermal effects, their tendency to contract in cold

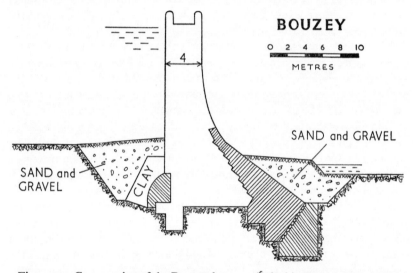

Figure 30 Cross-section of the Bouzey dam near Épinal in France. The shaded portion of masonry, the sand and gravel, and the clay were added between 1884 and 1889.

weather being liable to open up vertical cracks in the planes of their cross-sections. The Habra and Bouzey dams are among the earliest in which thermal effects were observed, and this affirmed the benefits of curving gravity dams, because curvature significantly reduces any tendency for temperature changes to open up cracks.

The Bouzey dam was undoubtedly weakened by its poor foundations, which allowed bodily movements, and by thermal expansions and contractions. Its actual failure in 1895, however, is nowadays attributed to the fact that the middle third rule was violated. This allowed horizontal tension cracks to develop on the water face into which water found its way, so aggravating an already dangerous condition by the development of uplift forces.

The Bouzey dam's failure is an excellent example of the old dictum that more is learnt from one unsuccessful structure than from dozens of successful ones. Quite apart from re-emphasising the need to build dams on solid rock foundations, it focused attention on several other points. Firstly, it confirmed that the design criteria of de Sazilly, Delocre and Rankine were basically sound, and that in particular the middle third rule had to be applied rigorously; it was fatal to allow tensile stresses to develop. Stemming from this was the first serious theoretical work on uplift pressures, either under a dam or in the body of the structure itself. Early theories due to writers such as Maurice Lévy in 1895 and Lieckfeldt in 1898 presupposed the presence of cracks into which water could penetrate, thereby introducing a vertical upward force which would assist the horizontal water load in trying to overturn the dam. Their solutions[17] consisted in effect of simply increasing the dam's weight to such an extent that the uplift forces would be balanced if and when they occurred. Lévy's work on these lines and that of several workers who followed him became well known and for many years was widely used. Subsequently, however, theories of uplift based on the idea of pore-water pressures were put forward, and these continue to be the subject of research and discussion. The earlier crack theories have now been abandoned. It is worth adding also that in France 'Lévy's uplift rule' was never intended to be mandatory in dam design. Its frequent application was in fact the result of a misinterpretation of a French ministerial circular of 1897 which covered another aspect of Lévy's work dealing with the determination of maximum stresses.

The failures of the Habra and Bouzey dams are not now regarded as having been primarily shear failures. Nevertheless the initial effects of tension and uplift very probably led to a shear failure in the final moments of collapse, and the considerable discussion of the two disasters which can be found in the engineering literature of the period shows that for the first time engineers began to ponder the magnitude of shear stresses in dams and the nature of their distribution.

While the Bouzey and Habra dams were not the only ones to fail in the second half of the nineteenth century, the Bouzey dam's fate was the most serious to befall a masonry dam and a structure, what is more, which was thought to have been rationally designed, a product of the newly developed theoretical climate which prevailed. In fact the failure served to strengthen the feeling of mistrust of existing methods of dam design which was already prevalent in some quarters. This concern had already manifested itself in at least two structures whose designs antedate the Bouzey disaster.

The Gileppe dam in Belgium was built between 1870 and 1875 with a profile greatly in excess of that required by the then current design methods. Evidently the engineers in charge of the work were acutely aware of the serious consequences which would attend a failure, and they designed a structure which was certain to be absolutely safe. Moreover, conscious of the fact that there were no sound theoretical means to allow for uplift effects, the specific gravity of the masonry was assumed to be 1·3 rather than 2·3, its actual value; this was a somewhat empirical approach but revealing just the same.

The Vyrnwy dam (Pl. 55), the first large masonry dam in Great Britain, was built between 1881 and 1892 for the water supply of Liverpool, the first occasion on which water was impounded in large quantities behind a dam on Welsh soil for use in England. At the time it was the largest reservoir in Europe, the dam itself being 1,350 feet long and 136 feet high. The dam has a notably more substantial profile than most of its contemporaries. Generally speaking, this was the result of a desire to ensure absolute safety—something which the designers were not convinced that the existing methods of dam design would guarantee. As one of them, Mr G. F. Deacon, remarked in 1896,[18] 'In 1881 [the year when the designs were first prepared] there was probably no high masonry dam in Europe so far watertight that an English engineer would take credit for its construction.'

It was this concern about the effects of water percolating through a dam that accounts, to some extent, for the Vyrnwy structure's being so adequately proportioned; it also explains why the dam was constructed in masonry to exceedingly high standards and why it was fitted with a network of drainage tunnels, apparently the first dam to feature such equipment.

By 1900 the end of the first phase in gravity dam theory was in sight. It was becoming increasingly apparent that existing design techniques were sound so far as they went and within the limits of the assumptions on which they were based, but that they did not go far enough. It was recognised, for instance, that the assumption of linear stress distribution, the basis of the classical method, was not valid, and by 1900 a number of authorities had drawn attention to this. Nor was there universal acceptance of the notion that one slice of a dam was sufficient for a general stress analysis of the whole structure. It was realised that a short, high dam such as Furens was essentially a different problem from a long, low structure such as Bouzey. Moreover, the failures of the Habra and Bouzey dams had shown all too clearly that horizontal water pressure was not the only important loading on a dam; uplift forces and thermal effects could be significant as well. There was a growing feeling, then, that the determination of the true state of stress in a dam required a much more elaborate treatment than had so far been undertaken. Finally, questions were being asked of the properties of masonry itself. Did it really result in a structure which was elastic, isotropic and homogeneous, and what precisely were its properties, notably its crushing strength? It was pointed out that the best of theories was useless if the material did not comply with the assumptions made about its properties.

In short the whole field of dam theory and design was ripe for the new developments which came at the beginning of this century.

While the gravity dam was the centre of theoretical attention between 1850 and 1900, some efforts were made to analyse arch dams, and several were built. Delocre himself, as a result of his advocating curvature in gravity dams, made a simple approach to the problem,[19] mainly with the idea of determining the extent to which arch action could be relied upon to develop in a curved dam. He concluded that it would so long as the thickness was less than one-third of the dam's water-face radius. M. Pelletrau, a later writer (1877), suggested that half the radius was the criterion, while J. B. Krantz was of the opinion that the radius must be as small as 20 metres (66 ft) for arch action to be effective.

All these early enquiries were based on what is known as the cylinder formula, the assumption being that an arch dam could be treated as a part of a complete cylinder, and that the dam's thickness at any depth was given by the well-known formula $t = \dfrac{pR}{\sigma}$. It is interesting that in Europe, while so much attention was focused on theories of gravity dams, these pioneer analyses of arch dams were given little attention. It remains something of a mystery that, in spite of the theoretical work which was done and even though thin arch dams already stood in Italy, France and Spain, the matter was dropped in Europe in favour of gravity and earth dams. And yet one French-built arch dam of the late nineteenth century still stands in Central America.[20] In 1888, an arch dam 11·6 metres (38 ft) high, with a mean thickness of 2·35 metres (7½ ft), was built on the Rio Grande as part of the control works of the first and ill-fated attempt to construct a Panama Canal.

While the arch dam faded into the background in Europe, it prospered in other places. Initially in North America, and near the end of the century in Australia, arch dams were built in some numbers. The first structure of this type to be built in the United States has become well known if only because it was one of the boldest examples of all, before or since. Indeed, for its time it was so slender that one is bound to state that the designer, F. E. Brown, was more daring than the contemporary state of knowledge really warranted.

The Bear Valley dam[21] was built in 1884 to impound irrigation water in the San Bernardino mountains in California. The site was a remote one, necessitating the haulage of all equipment, tools and materials over seventy miles of rough mountain and desert roads. Brown therefore elected to build as thin an arch dam as he dared, in order to reduce the transportation problem to a minimum. Funds for the work were limited as well,

although the benefits derived from the completed structure ultimately
proved its value. The Bear Valley dam was designed by the cylinder
formula and constructed of roughly cut masonry blocks, covering a
rubble masonry core set in Portland cement mortar. An indication of the
then current diversity of thinking about design stresses can be drawn from
the fact that the Bear Valley dam was designed for a maximum stress of
620 lb/in.², very much higher than the European figure for gravity dams.

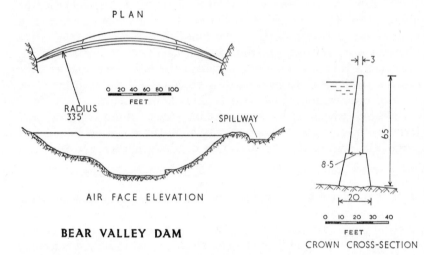

Figure 31   F. E. Brown's dam of 1884 in the San Bernardino mountains of
California.

The structure stood to a height of 65 feet with a radius to its vertical
air face of 335 feet. The thickness was 3 feet at the crest increasing to $8\frac{1}{2}$
feet near the base. Figure 31 should leave no doubt as to the slenderness
of this amazing piece of work.

From the outset numerous people expressed concern for the Bear
Valley dam's safety. Quite apart from being so thin, its spillway was
dangerously small, and it was thought likely that a severe flood would
overload the structure and break it all too easily. However for twenty-six
years the structure performed without trouble and was finally replaced
by a multiple-arch dam in 1910. It was the pressing need for more water
in and around Redlands, especially for the citrus groves, which prompted
the construction of the new dam. The original Bear Valley dam now
stands in the depths of the enlarged reservoir.

The Bear Valley dam was the first of a number of arch and arched dams
in the United States, and the fact that there were plans in the 1890s to
carry out deflection measurements on the Bear Valley dam suggests a
degree of interest in the behaviour of such structures which was not to

be found elsewhere. Brown, the Bear Valley dam's designer, was engaged in 1886 to plan an irrigation and water-supply dam on the Sweetwater river for the benefit of San Diego and National City. His design was nearly as bold as before, but this time the backers of the scheme lost their nerve, and the Sweetwater dam when finished in 1888 was a much more substantial structure than originally intended.[22] It is an arch dam nonetheless: 98 feet high, 12 feet thick at the crest, 45 feet thick at the base and curved to a radius of 222 feet at the top.

The Sweetwater dam was most carefully constructed from locally quarried porphyry, a very heavy stone, and imported English and German cement. The quality of the dam's construction was severely tested in 1895, when heavy rainfall flooded the reservoir and for nearly two days the dam's crest was submerged to a depth of 22 inches (Pl. 56). The flood played havoc with the bed-rock below the dam—no less than 10,000 cubic yards of rock were washed away—and part of the outlet works was destroyed. But the dam experienced no damage, and its survival appears to have helped convince American engineers that a well-made arch dam was a perfectly sound proposition. It was also concluded, as it already had been in Europe, that even in a gravity dam curvature would increase a dam's stability and provide a degree of resistance against the formation of tension cracks caused by temperature variations. Thus the Hemet, La Grange and San Mateo dams, all built in California between 1887 and 1895, were arched dams, and more arched dams were built in other states.[23]

In March 1909 L. A. B. Wade presented a paper to a meeting of the Institution of Civil Engineers in London entitled *Concrete and Masonry Dam Construction in New South Wales*.[24] Both the paper and the discussion which followed are of more than usual interest. It emerges that one of the earliest dams in Australia, the Parramatta dam, built to a height of 41 feet in 1852, was a masonry arch dam of 160-foot radius. In the 1890s nine more arch dams were built in New South Wales, followed by three others by 1906. All the dams were, for the time, extremely slender, and this fact caused something of a stir at the meeting of 1909, especially because for so many years gravity dams and their relatively heavy profiles had dominated the Institution's discussions of dam design.[25] For Sir Alexander Binnie, the collection of Australian dam profiles displayed produced a 'blood-curdling sensation', while C. E. Jones was extremely critical and warned that in his view Australian engineers had taken dam-building to a dangerous point. What exactly was it about the dams that caused these and other similar feelings of alarm and concern? We cannot look at all the dams in detail (a selection, however, is shown in Fig. 32), but leaving aside the much earlier Parramatta dam, it can be said that of the other dozen their height-to-mean-thickness ratios range from 2 up to 10·4,

H

while for seven of them it is 5 or greater. The thinnest dam, Medlow (65 feet high), is 9 feet thick at the base, and the tallest one, Lithgow No. 2 (87 feet high), is 3 feet thick at the crest and 24 feet at the base.

Cost was the basic reason for the choice of thin arch dams, simply because so much less material was required for an arch dam than for the equivalent gravity dam. It was also on the grounds of expense that all the structures were made of Portland cement concrete for which a crushing strength of as high as 20 ton/ft^2 (310 lb/in.^{2}) was allowed in six of the dams. In some cases, however, the compressive strength of the abutment rock determined the allowable stresses in the dams themselves.

The dams were all designed with nothing more than the cylinder formula $t = \dfrac{pR}{\sigma}$, and it was remarked at the time that this was a far from adequate means of analysing or designing an arch dam. It was essentially the fact that high compressive stresses (compared with what was usual in Europe) had been used in conjunction with an over-simplified theory that caused people's fears to be aroused. Nevertheless these Australian arch dams had made their point. They had all been built and all performed satisfactorily and safely. True, some of them produced cracks due to temperature changes, but in such thin profiles even arch dams were prone to this weakness, especially if they were relatively long; it was only in those of high length/height ratio that cracking occurred.

Whatever the weaknesses of the cylinder theory and despite the apparently high stresses allowed, the thin arch dams of New South Wales were a success. They showed that arch dams of unprecedented slenderness were perfectly feasible. As a result it became the responsibility of theoreticians to confirm and explain on paper what had been forcibly demonstrated in practice.

The dozen arch dams of New South Wales and some others in Australia are a reminder of another development which was taking place at the end of the nineteenth century: the use of concrete as a material of construction. Its partial use for the cores of masonry dams and the core walls of earth dams goes back at least to the 1870s. Concrete formed the core of the Boyd's Corner dam built to supply water to New York in 1872, and a year earlier the Lynde Brook earth dam in Massachusetts was built with a concrete core wall. The San Mateo dam previously mentioned, built between 1887 and 1889 to supply water to San Francisco, was the first dam to be made entirely of concrete. It was used simply because suitable masonry was not available locally and would have proved expensive to bring in from elsewhere. It should also be remarked that this first all-concrete dam was a very large structure—170 feet high and 680 feet long. The task of building the San Mateo dam was itself an important piece of engineering because the techniques required to place a 'wet' material

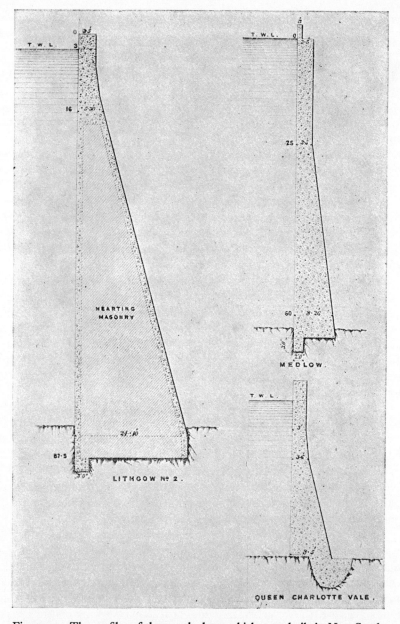

Figure 32 The profiles of three arch dams which were built in New South Wales around 1900. These thin cross-sections caused something of a stir when they were discussed in London in 1909. (From *Min. Proc. Instn. Civ. Engrs.*, Vol. clxxviii.)

A History of Dams

rather than a 'dry' one on such a large scale posed new problems.[26] The work must have been well executed, however, because in 1906 the San Mateo dam successfully survived the San Francisco earth-quake.

In 1899 the use of concrete in dam-building was taken a stage further in another Californian project. The Upper Otay dam, built between 1899 and 1900 to supply water to San Diego, is a thin arch dam 84 feet high, 4 feet thick at the top and 14 feet at the base. Its strength was increased by the lavish use of iron wires at the base, the first example, it would appear, of a concrete dam being reinforced.

In Great Britain concrete found its way only slowly into dam-building. Some early examples are the Blackbrook dam near Loughborough, built in 1906; the Blackwater dam, 86 feet high and over 3,000 feet long, built in 1909 for hydro-electric work in Scotland; and the Alwen dam of 1911–16 which supplies water to Birkenhead.

British engineers had, however, made use of concrete construction before 1900 in one of their overseas projects. The dams built by British engineers in the 1890s in India were among the largest in the world, and in the case of the Periyar dam, mentioned earlier in connection with Professor Rankine's theoretical work, the problem was to build a structure 160 feet high in the absence of suitably skilled masons. This difficulty was overcome by the use of concrete, which was used throughout, except for a thin facing of masonry.

Earth dams continued to be built in large numbers in the second half of the nineteenth century but without the benefit of any theoretical studies. The percentage of earth dams which experienced a partial or total failure was alarmingly high, and this clearly showed that the various empirical design rules being followed were by no means all reliable. Numerous published lists indicate that some tens of earth dams in various parts of the world experienced some sort of serious failure before 1900, and the majority of these were in the United States.[27]

By far the most common cause of these failures was insufficient spill-way capacity—the South Fork dam failure of 1889 was such a case—a defect which stemmed from a continuing inability to predict accurately the volume of overflow with which dams would have to cope. While masonry and concrete dams often survived the consequences of this ignorance, e.g. the Sweetwater dam, earth dams were by contrast very vulnerable, and this aspect of their operation was a stimulus to research into the hydrology of rivers and catchment areas.

The two other predominant causes of failure were piping and slips in the embankments themselves. Before 1900 the percolation of water through and under earth embankments and the stability of earth dam slopes were problems about which very little was known. The continuing poor performance of earth dams eventually put pressure on the engineering profession to take up these matters in the twentieth century.

It was mentioned earlier that in Great Britain the earth dam has enjoyed a better safety record than in most other countries. The fact remains, however, that a few failures have occurred, and one was the worst dam disaster ever experienced in Great Britain.

In 1850 the Woodhead dam, the one furthest upstream of the series in the Longdendale valley, collapsed, and while there is no certainty as to the cause of failure it seems most likely that water percolating through the grits beneath the dam undermined it. The Holmfirth dam five miles south of Huddersfield was built on similar foundations and in 1852 met a similar fate.

The crucial year, however, was 1864. The Dale Dyke dam was built in 1858 to supply water to Sheffield. It stood to a greatest height of 95 feet and was 1,250 feet long. Its thickness at the crest was 12 feet and at the base in the centre of the valley it was 500 feet thick; the slopes of the two faces were 1 in 2½. The core wall was made of puddled clay and was 4 feet thick at the top, 16 feet thick at the base of the embankment, and below this level extended 60 feet into the foundations at some points. The upstream and downstream parts of the embankment were composed of a mixture of shale and rubble excavated from the bed of the reservoir. These materials were dumped loosely into place without proper compaction and the dam's water face was covered with a layer of rough masonry blocks. The outlet consisted of twin 18-inch cast-iron pipes running obliquely through the base of the embankment in a trench filled with puddled clay. The dam had an overflow spillway 24 feet wide and 11 feet deep.

In the spring of 1864 the outlet valves in the Dale Dyke dam were closed in order to raise the water level in what is often called the Bradfield reservoir. This was the first time it had been filled right up, and it was to be the last. On the afternoon of 11 March Mr John Gunson, an engineer of the Sheffield Waterworks and the man in charge of the dam, went to inspect the full reservoir and found that all was well. The situation changed rapidly for the worse, however. In the evening, as a Mr Hammerton crossed the dam, he noticed a crack in the air face. This gradually widened and at 11.30 p.m. the dam give way. Nearly a quarter of the embankment near the middle collapsed, and an estimated 200 million gallons (730 acre-feet) of water poured out of the reservoir. A photograph of the breached embankment is shown in Plate 57.

The official report[28] on the disaster, published in May 1864, states that the wave of water released carried 40,000 cubic feet of water per second and bore down on Sheffield, seven miles away, at a speed of 18 miles per hour. Between the dam site and Owlerton the flood destroyed everything in its path because it was confined to a narrow valley. Mercifully the valley opens out beyond Owlerton, and this allowed the water to disperse sideways, thereby saving Sheffield from a worse fate. Even so, some parts of the city were flooded to a depth of nine feet, and large areas were left covered with a thick layer of wood, mud, sand and stones when the waters subsided.

The failure of the Dale Dyke dam killed 250 people, destroyed 798 houses and seriously flooded more than 4,000 others. Dozens of other properties—mills, factories, shops and workshops—were totally or partially destroyed, and the official report even details how many horses, cows, donkeys, pigs and other animals were drowned.

As far as the dam failure itself is concerned, the results of the enquiry were inconclusive, although it was generally agreed that the structure was a very poor piece of work; the core wall was too thin, the embankment was much too loosely built of the wrong materials, the overflow was totally inadequate for the floods likely to occur, and the outlet pipes were laid incorrectly and without proper support.

In retrospect it seems likely that the dam's failure was initiated by the faulty positioning and laying of the outlet pipes. As the heavy bank settled the pipes were probably displaced, and water percolating through the pervious embankment began to wash away the puddled clay surrounding them. This undermined the core wall at the point where the pipes passed through it, and very soon a considerable volume of water was eroding away the interior of the dam. It is conceivable that at this point the central portion of the dam subsided enough to allow the already full reservoir to spill over the poorly constructed crest. Thus the centre of the dam was eroded from above and below and a collapse was inevitable.

At the time, the accident made a considerable impression in and around Sheffield as can easily be imagined; Figure 33 shows a typically Victorian reaction to the event. On a more serious note, however, it is interesting that in the official report a recommendation was made to the effect that all dams and reservoirs in Great Britain should, by Act of Parliament, be subject to 'frequent, sufficient and regular' inspections. This suggestion, however, was not taken up, and it took the Dolgarrog disaster of 1925 to bring about the Reservoirs (Safety Provisions) Act of 1930.

In Great Britain the Bradfield disaster brought home the fact that the dam-builder has a considerable social responsibility. Well executed, his work is of great benefit to the community, but if it is not, a dam failure is perhaps the most serious man-made catastrophe likely to occur in peacetime. The rapidly growing literature on dams in the latter decades of the nineteenth century makes increasingly more frequent reference to this point, particularly because so many big dams were being built near

A COPY OF VERSES

Written on the Sad and Awful Calamity that happened by the

BURSTING OF BRADFIELD RESEVOIR,

NEAR SHEFFIELD,

When upwards of 240 Human Beings were swept into Eternity, besides a great loss of Property.

You feeling christian's both far and near,
To these few lines lend a silent ear;
A sad calamity I will make known,
That lately happened near to Sheffield town.

Sheffield town with water to supply,
They at Bradfield formed a large Reservoir;
But the bank gave way, and for miles around,
Men, women, and children have been drowned.

Both mills and cottages how sad to say,
From their foundations were washed away;
While sisters, brothers, husbands, and wives,
In the mighty flood lost their precious lives.

The angry water in the midnight hour,
Both life and property it did devour;
Whole streets of cottages were washed away,
With their sleeping inmates as in bed they lay.

That fatal morning when daylight came,
To describe the scene was all in vain;
Workshops and houses did in ruins lie,
Caused by the bursting of the Reservoir.

Ah! little did those poor sufferers knew,
That fatal night as they to bed did go;
Such dreadful danger did them await,
Or that they'd meet such a dreadful fate.

Mothers, and fathers, in anguish wild,
Lament's the loss of their darling child,
The sister cries with a bitter tear,
And the brother weeps for his sister dear.

No pen can write, nor no tongue express,
The dreadful suffering and sad distress,
Some scores were crushed beneath stones and mud,
And lost their lives by the dreadful find.

This sad and fearful catastrophe,
For years to come will remembered be;
Where so many people were snatch'd away,
To sleep in death till the Judgment day.

Bebbington, Printer, Manchester.

Figure 33　A copy of a set of verses composed at the time of the Dale Dyke catastrophe and circulated as a broadsheet. (Courtesy of Sheffield City Libraries.)

cities in an effort to meet the ever-growing need for more domestic and industrial water.

Other socially oriented aspects of dam-building were emerging as well. The creation of Lake Vyrnwy, for example, involved the drowning of a whole village if the best available site was to be utilised. The loss of land incurred by forming a large reservoir had its obvious social and economic consequences to which some thought had to be given. When Manchester was obliged to go further afield for its water supplies—the Longdendale reservoirs were being worked to their maximum by the 1870s—it selected Haweswater and Lake Thirlmere as the most suitable sources. Manchester Corporation met tremendous opposition from landowners and water authorities, and the schemes were only approved at the expense of making considerable payments and concessions to these various interests.[29] Moreover the Thirlmere scheme was an early example of opposition to dams on aesthetic grounds, it being claimed that such structures and their ancillary equipment defaced the landscape. Dam-building around 1900 had reached the stage where many aspects apart from the purely technical must be taken into account, and in fact the non-technical problems have often been equally difficult to solve.

The last development in the nineteenth century which we have to consider has turned out to be one of the most important. Indeed, it is among the most significant technological consequences of the whole of nineteenth-century science and engineering.

During the nineteenth century various people had gradually advanced the development of the water-turbine; in its applications it was the logical successor to the water-wheel, but in its mode of operation, particularly the reaction or pressure turbine, it was a fundamentally new idea whose progress was closely linked with an improving understanding of hydro-dynamics.[30] The most important names in the development of this new prime mover were mainly French and American: Fourneyron, Jonval and Girard in France, Pelton, Howd and Francis in the United States. They produced water-driven machines which in terms of high performance and small size were a revelation, particularly in comparison with the recipro-cating steam engines which at the time were so widely used.

Among the major scientific discoveries in the early part of the century, those connected with electro-magnetism had profound results for technology.[31] The development of electric generators was among these, and one feature of electric power in particular was of supreme significance, namely that it is the only form of energy in a ready-to-use state which can be transmitted over long distances. Exactly how this should be done, especially whether it was alternating or direct current which offered the better method, was at first far from clear. But it was absolutely clear that the whole question of utilising hydro-power resources was on the verge of major changes. For the first time it was possible to develop power from water at one place and then transmit it for use somewhere else, perhaps in a town or city tens or even hundreds of miles away.

In order to drive a water-turbine a head of water is required, and a number of early hydro-electric installations made use of natural differences in water-level such as the famous and epoch-making Niagara Falls project[32] which was built in the 1890s to supply electric power to Buffalo, twenty-six miles away. Nature does not provide such natural sites to order, however, and it was inevitable that dams would be called upon to provide a head of water most of the time. Thus, three separate technologies—those concerning dams, water-turbines and electric generators—experienced a marriage towards the end of the nineteenth century. It is one of the supreme examples of seemingly diverse branches of engineering coming together to found a new branch of the subject, one which in this case has been and is still of world-wide importance.

Initially dams were used for hydro-electric installations as part of run-of-river schemes. That is to say, the dam merely served to raise the level of a river but did not create any significant degree of storage. The first example of such an arrangement was built in 1882 on the Fox river at Appleton, Wisconsin, and for twenty years or so it met the small requirements of the locality. The Wilamette Falls scheme at Portland, Oregon was initially (1889) based on a natural drop in water level, but in 1894 a dam was added to provide more control and a steadier supply to the turbines.

It was in the western United States that conditions were especially favourable to the development of hydro-power schemes. In various mountainous regions the rapid fall of the rivers favoured the establishment of high heads of water, and the water itself, once past the turbines, could then be used for irrigation.

The Colorado river was dammed for hydro-electric generation in 1893, near Austin in Texas.[33] The dam, 65 feet high and nearly 1,300 feet long, was the first large masonry dam (in fact it had a concrete core) ever built for a hydro-electric scheme, and it supplied generators whose total output was nearly 15,000 h.p., a very high figure for the time. But the project was short-lived. On 7 April 1900 heavy rain caused a rapid and large rise in the Colorado's flow, and soon a sheet of water eleven feet deep was rushing over the dam. The central section gave way and slid bodily downstream, stopping some thirty feet from its original position. This is an instance of a dam failing by sliding, and the accident was subsequently agreed to have been caused by erosion of the dam's downstream toe, coupled with the fact that the foundations were poorly constructed.

In the case of the Austin dam siltation appears on the scene again. The reservoir was very large (53,500 acre-feet) and by 1897 measurements showed that 41·5 per cent of this volume had been taken over by sediment. This appears to have indicated to American engineers that if a reservoir was to have a usefully long working life, its capacity would need to be calculated in relation not only to the annual discharge of the river but also to the amount of silt carried.

Near the turn of the century hydro-electric dams were built at Massena, New York and Ogden, Utah; the former featured a power-house (the installed capacity in 1902 was 35,000 h.p.) integral with the dam, while the latter, also used for irrigation, included some reinforced concrete in its construction. These structures were contemporary with the earliest hydro-electric dams in Europe where France and Italy led the way. In France the dam of Avignonet was begun in 1899 to generate power for Grenoble, and the Sioule and Miodeix dams were completed soon after in the Department of Puy-de-Dôme. In Italy the first hydro-electric dam was built between 1880 and 1883 near Genoa, and its construction heralds the dawn of Italy's entry into dam-building in the twentieth century, a development due in large measure to that country's great hydro-power potential.[34]

The key structure in the whole of this chapter is the Furens dam about which Edward Wegmann made the following statement in 1888: 'The Furens dam, which surpasses all other reservoir walls in height, has a maximum elevation of 194 feet above its foundation. Higher dams may be built in the future, but will probably be exceptional.'[35]

Wegmann was wrong. The rationally designed high dams of the nineteenth century were merely a beginning, a prelude to the twentieth-century developments which have placed dams of the size of Furens in the ranks of the unexceptional. This, however, is not to deny the importance of the earlier period; the second half of the nineteenth century was a crucial phase in the story, full of significant advances.

10

The Twentieth Century

SINCE 1900 FAR more dams have been built than were already standing at the turn of the century. All over the world this increase in numbers has proceeded hand-in-hand with a steady increase in size as all the basic types of dams—arch, gravity, buttress, earth and rock —have been applied more and more to the water-supply, power, irrigation and river-control problems of virtually every nation. At the same time, however, dam-building has to a large degree been standardised, and twentieth-century dams in many ways lack the individuality of the works of earlier centuries. Economic demands and improved techniques of design and construction have forced and also allowed dam-building to become a largely routine affair for much of the time. In this chapter it is intended to trace the significant developments and major trends which underlie this situation and to mention the exceptional structures which have in some way been pioneers or represent important stages in the story.

As far as the stress analysis of dams is concerned, the twentieth century opened with a major skirmish. In 1904 the second in the series of *Drapers' Company Research Memoirs* was published;[1] it contained the results of some theoretical and experimental work on gravity dams carried out at University College, London, by L. W. Atcherley and Professor Karl Pearson. The conclusions they reached were essentially these: it was not sufficient to design dams on the basis of calculations applied to horizontal sections, and in fact it was across vertical sections that dangerously high tensile stresses would develop first; secondly, the distribution of shear stresses must be more nearly parabolic than linear, and the maximum intensity of shear was much higher than had previously been supposed. Atcherley and Pearson also questioned (as others had done earlier) the validity of assuming that a gravity dam could be treated as a series of disconnected slices.

Atcherley and Pearson claimed to have verified their analytical work by means of tests on models, apparently the first occasion when this technique was applied to dams. But their experimental profiles were made simply of strips of wood and were not at all suitable as a means of simulating the behaviour of a dam. Nevertheless a startling new theory was proposed, its effect on the engineering world being the more dramatic because Atcherley and Pearson advanced the view that, even though all existing dams might be safe with regard to vertical direct stresses, they were also subject to appreciable horizontal tension and excessive shear.

The controversy sparked off by the above research lasted for four years. It was an interesting episode because, quite apart from the fact that the theory of dams was advanced as a result, the principal protagonists— Professors Karl Pearson, W. C. Unwin and others—carried on their debate in the pages of *Engineering* and other publications in full view of the engineering profession at large. A thorough and vivid account of the controversy was published in 1950 by Professor A. J. S. Pippard.[2] In 1908 the affair was effectively wound up by the experimental researches of J. S. Wilson and W. Gore.[3] Using india-rubber models of gravity dam profiles, loaded in such a way that stresses due to both dead and live loads were properly reproduced, they succeeded in showing to the satisfaction of most engineers that the current methods of designing gravity dams were sound, and that in a dam designed by the classical theory there would be no tension except in a small area near the water-face base corner— the heel of the dam. The experiments of Wilson and Gore caused the earlier alarmist theories of Atcherley and Pearson to be rejected. It was also confirmed that, although the classical theories were safe as a design method, they did not yield an exact stress analysis, and that this could only be obtained through the equations of the theory of elasticity.

The first application[4] of the equations of elasticity appears to have been made in 1898 by Maurice Lévy, better known for his work on uplift. In 1913 in Germany, Fillunger extended the solution to take dead loads as well as live loads into account. From the outset the theory of elasticity was applied not just to a dam profile but to this in conjunction with a section of the foundations. Assuming that both dam and foundation were isotropic, homogeneous and perfectly elastic, numerous mathematicians endeavoured to stress analyse the two as a continuous system, their particular concern, of course, being that it was in the region of a dam's base that the most serious errors in the classical theory occurred. L. F. Richardson in 1910 carried out a most elaborate solution (and one of the best) using Airy stress functions, and produced results substantially in agreement with Wilson and Gore's model tests. Other attempts based on the stress function were due to Karl Wolf in Austria and J. H. A. Bratz in the United States; the latter included in his work an attack on the problem by means of photo-elasticity.

One is bound to point out that many of these applications of elastic theory, excellent though they were as mathematical exercises, made little immediate impact on dam-building, and the design engineer continued for the most part to depend on the classical theory. A paper which can now be seen to have signalled the beginning of a new phase in the application of the theory of elasticity was by O. C. Zienkiewicz in 1947.[5] In this case a solution was arrived at using Southwell's method of relaxation, and the results obtained are particularly interesting because more than any others they confirmed the conclusions reached forty years before with rubber models. Zienkiewicz's work further substantiated the validity of the classical theory as a useful design tool, and it continues to be used.

One of the reasons why the Pearson–Unwin controversy of 1904–8 generated so much interest was because it coincided with proposals to heighten the Aswan dam. The first dam ever built across the Nile's mainstream was completed in 1902, one of the finest dam-building achievements of all time.[6] The structure was made of locally quarried granite and stood to the unprecedented length, for a dam of its type, of 6,400 feet. In all it was fitted with 180 low-level sluices with which to control the river and feed water to the Assiut barrage, an irrigation dam 350 miles downstream and of interest as the first dam whose construction benefited from the research on seepage which had been carried out in India near the end of the nineteenth century.

The Aswan–Assiut scheme gave such a boost to Egypt's prosperity that in 1905 it was decided to increase the Aswan dam's height from 65½ to 88½ feet, thereby more than doubling the reservoir's capacity. The concern aroused in London, however, by the work of Atcherley and Pearson threatened at one stage to prevent this addition, and work was in fact halted for a time to consider the situation more carefully. Sir Benjamin Baker, the engineer in charge at Aswan, soon concluded that there was really no reason not to continue, and the heightening was finished in 1912. The demand for irrigation water continued to grow, however, and between 1929 and 1933 the Aswan dam was heightened yet again, this time by 29½ feet.

The original plans for Nile dams at Aswan and Assiut were drawn up by Sir William Willcocks. Following his work in Egypt, he was invited by the Turkish government to visit Iraq to survey the problem of revitalising the agriculture of a country which for six centuries had lain waste, a sad contrast with the days when the Tigris and Euphrates had sustained such splendid irrigation systems. Willcocks' work in Iraq was remarkable. In two years, 1908–10, he surveyed the whole country and its water resources and drew up plans to utilise the 'two rivers' once more. He also found time to examine Iraq's old irrigation systems, and it is of some interest that his own schemes were in no small measure shaped by the achievements of his ancient predecessors.[7]

The first major control dam specified by Willcocks was the Hindiya barrage across the Euphrates, completed in 1913. The subsequent development of Iraq's irrigation and flood-control works have been based very largely on Willcocks' original plans even though they were not realised in his lifetime. The 1,600-foot long Kut barrage was built between 1937 and 1939 as the basic element in the control of the Tigris, and this was followed in the 1950s by the construction of diversion dams at Ramadi on the Euphrates and Samarra on the Tigris.

All the irrigation works built by British engineers in Egypt, India and Iraq early in the twentieth century featured either gravity or earth dams; and with rare exceptions this was the pattern in Great Britain as well. It was principally in the United States and Europe that the design and construction of arch dams underwent the bulk of their development in the twentieth century. What is especially notable is the emergence of two distinct schools of arch dam thought; it is particularly apparent in the history of methods of designing and analysing arch dams.[8]

The first two large arch dams built in the United States in this century were both planned in 1905 to form irrigation reservoirs in Wyoming. The Pathfinder dam was completed in 1907 and the Shoshone (now Buffalo Bill) dam in 1910, the former to a height of 214 feet and the latter to a height of 325 feet. Both stand in deep and narrow gorges of granite, and when it was built the Shoshone dam was the highest in the world. The procedure by which both dams were designed represented a major advance.

As we have seen already, the first step in visualising the action of gravity and arch dams is to regard them as sets of separate elements: vertical cantilevers in the one case, horizontal rings or arches in the other. It was in connection with the design of the Pathfinder and Shoshone dams that the basis of a new method of arch-dam analysis was laid down. It is a method which, in essence, considers the action of cantilevers and arches simultaneously. It is assumed that an arch dam can be regarded as a co-existent system of independent horizontal arches and vertical cantilevers, the arches being fixed at their ends and the cantilevers being fixed at the base and free at the top. The water load acting on the arch dam is then distributed between the two structural systems in such a way that the movements of both arches and cantilevers are compatible at their points of intersection.[9]

The version of this technique used in 1905 was an elementary one. The radial deflection of each arch at its crown was computed by cylinder theory, and those of a single cantilever, the one at the crown, by standard beam theory. By trial-and-error methods, the water load was divided between the two systems in such a way that the arch deflections were equal to the crown cantilever deflections at each point of intersection. The so-called crown cantilever method assumes, therefore, that the load distribution at the crown is applicable at all other points around the dam.

Conceptually, this method of analysis was a considerable advance and it founded a whole school of American thought on arch-dam design. In 1929 the procedure was improved by introducing elastic theory to determine the arch deflections in conjunction with a series of cantilevers rather than just one. Trial-and-error methods were once more used to adjust the radial deflections, and it was on this basis that the Gibson dam in Montana was designed in 1926. In the 1930s the 'trial-load method', as it is generally known, was extended even further in order to cope with the design of the Hoover dam, a giant structure which put all previous dams into a decidedly lower category on virtually every count. Standing to a height of 727 feet, it is still among the highest in the world and remains one of the finest pieces of dam-building on record (Pl. 58).

In its final form the trial-load analysis was elaborated so as to equalise not only radial deflections but also tangential deflections and twist deformations at each point of intersection of the arch and cantilever elements. If temperature effects and foundation deformations are taken into account as well, the full trial-load analysis of an arch dam is an extremely lengthy piece of work by manual computation, although it has been done on several occasions.

In Europe the trial-load method never became popular, which is not to say that it was never used at all. Two factors in particular seem to account for this state of affairs. There is first of all the point about the enormous amount of computation involved. Quite apart from being tedious, the sheer length of time required can make the design of a dam very expensive, and to European engineers unnecessarily so. Secondly, a number of dams designed by the trial-load method turned out to be massive in the extreme. The Hoover dam was one example; another was the Hungry Horse dam. The essence of an arch dam, after all, is that it should resist the applied loads by arch action. There seems to have been a feeling in Europe that the trial-load method allowed gravity action to figure too prominently in the design process, with the result that the final shape was really that of an arched dam. It was the determination of European engineers to exploit arch action as much as possible that allowed them to develop arch dams to a point of great structural sophistication and marked economy of material. In this they have, of course, been aided by the large number of suitable sites at their disposal—deep, narrow gorges in strong rock.

European techniques of arch-dam design were characterised in the first half of the twentieth century by methods based on independent arch theory. In this case a dam is treated as a pile of separate arches fixed at their ends. In a sense it is the trial-load concept without cantilevers, and that basically is the point. The arches by definition are required to carry all the load by themselves. In Italy, France, Switzerland, Austria and other countries, various types of arch dam were evolved from independent arch methods, in every case the arches being treated each as separate structural elements. In this way the dam's thickness, radius of curvature, central arch angle and so on can all be varied at different levels in an effort to produce the most effective shape. In France André Coyne pioneered the concept of 'plunging arches'—a variation on the basic theme in which the dam is divided into a series of sloping arches rather than horizontal ones.[10]

Especially as a result of Italian work, the shape of arch dams gradually advanced from the single curvature types of the early decades of the twentieth century to the double curvature, or cupola, types of the post-war period.[11] Such structures represent some of the most elegant civil engineering of modern times. Cupola dams are nothing less than thin shells, and numerous procedures have been put forward to design them as such, notably by Tolke in 1938 and later on by Krall, Lombardi and Parme.

At first sight it is perhaps paradoxical that the arch dam in the twentieth century, despite being developed to such slender proportions utilising a minimum of material, has had a remarkably good record of safety. Very few arch dams have experienced failures and none, it seems, have collapsed as a result of the walls themselves being defective. The explanation of this can be traced to three things. The arch shape is inherently very sound for dams; it has considerably greater reserves of strength than the gravity dam, and the more it is loaded the tighter is it thrust against its supports. The strength of an arch dam is its shape; that of a gravity dam, its weight. Secondly, the very fact that the arch dam was intended to be slender and economical encouraged dam-builders to be cautious and thorough. Arch-dam design was the subject of far more careful and elaborate research than was the case for other types. Lastly, the materials of arch-dam construction (and those of gravity dams as well) were themselves greatly improved. The application of concrete to dam-building around 1900 provided engineers with a material of more reliable and predictable properties and one whose compressive strength was gradually increased.

The mention of safety records naturally leads to the topic of earth dams which, as we noted earlier, were very vulnerable in the nineteenth century, and so they continued to be early in the twentieth century; earth dams continued to suffer total or partial failures at the rate of several each year. It was vital that this serious state of affairs be looked into, even though the problems were recognised as being extremely difficult to solve. Indeed, it was probably because the mechanical properties of soil, and particularly of clay, were so much more obscure than those of the more straight-forward building materials—iron, steel, masonry and concrete—that people were discouraged from making a start. It was not at all easy to see *how* to start.

To be safe, an earth dam must satisfy four basic requirements: it must not be over-topped by a flood which will almost certainly cut right through it; seepage under the dam must not be allowed to undermine it; seepage through the embankment must not wash it away; and the slopes of the embankment must be so constructed that they will not slip.

From the outset the analysis of seepage effects was based on D'Arcy's law which was developed in the middle of the nineteenth century and first published in 1856. The earliest applications of this law to the flow of water through pervious media were confined to soil and sand, and this limitation also applied to dams. Following the experimental researches of Beresford and Clibborn[12] in India around 1900, Koenig[13] in 1911 tried to deal with seepage under masonry dams built on sand foundations with formulae worked out by Slichter for sand filters. Slichter's formula and others, notably one developed by Hazen, were subsequently applied to earth dams in conjunction with line-of-saturation concepts of the type proposed by J. D. Justin.[14] In the end, however, these, along with Bligh's creep theory of piping, gave way to more fundamental theories developed from soil mechanics.

As early as 1886 Forchheimer had demonstrated that the distribution of water pressure and velocity in a seepage medium was governed by Laplace's equation, but although methods of solving the equation were available in the early 1900s they were not at the time applied to earth dam problems.

In 1922 Karl Terzaghi published a correct explanation of the mechanism of piping, and three years later he showed that D'Arcy's law was also valid for seepage through clay. Terzaghi followed this up in the 1930s by drawing attention to the importance of the forces developed within earth dams by water percolating through them.

In 1937 Arthur Cassagrande published[15] his comprehensive paper 'Seepage through dams' which demonstrated that the D'Arcy and Laplace equations were the essential basis of seepage studies, and from his work all subsequent developments and refinements have stemmed.

It was also in the 1930s that the stability of earth-dam slopes was first investigated analytically. It will be noticed that this was nearly a century after Collin's work, and indeed it needed a rediscovery of the essence of his research by Pettersson, Bell and Terzaghi between 1915 and 1925 to get the study of cohesive soils back on the right lines. It was Pettersson who pointed out that the surface of failure of a clay slope is curved, and in 1926 Fellenius showed that for all practical purposes this curve is a circle.[16]

Early applications of the Swedish circle method to earth dams were inconclusive and led to a considerable amount of research and discussion. The fundamental difficulty was that theoretical predictions did not correspond with actual practice.[17] Papers presented at the First and Second World Congresses on Large Dams by Terzaghi (1933) and Mayer (1936) pointed out that a number of slips which had actually occurred could not have been predicted. Ultimately it was recognised that these discrepancies were due to a lack of knowledge of the shearing resistance of soils, and especially the influence of pore-water pressures. These questions have since been the subject of research, and at the same time methods of computation for the stability analyses themselves have been improved.

All the essential elements of earth-dam theory were worked out by about 1940 and their development and elaboration goes on. The trend towards higher and bigger earth dams has in part been an outcome of this theoretical work, but it has also been encouraged by the introduction of greatly improved methods of construction. It is most important that the materials used to build an earth dam be properly placed and compacted. The importance of proper compaction was not fully appreciated for some time, although it was often achieved to some extent in the normal process of transporting material on to the dam and laying it in place. The sheepsfoot roller came into use at the beginning of the twentieth century, and between 1930 and 1950 it was the standard compaction device. After 1950 the rubber-tyred roller began to replace other types because it was found to be more economical, and its use rapidly became standard. It has been recognised for some time now that artificial compaction of earth dams is an absolute necessity, theoretical work having indicated that compaction is itself subject to laws and can to a large extent be 'designed'.

The economics of earth-dam building have been progressively influenced in the twentieth century by the development of increasingly powerful earth-moving machines. It was principally this trend which allowed the construction of really large earth dams whose sheer volume could not have been handled in any other way, except perhaps by the hydraulic-fill technique.[18] The latter was commonly used in the United States for nearly a century, and the ultimate example of the process is the Fort Peck dam which contains 125 million cubic yards of material. A slip in this structure in 1938, however, seems to have been a factor in the abandonment of the technique and it has never been reintroduced.

The widespread application of rock-fill dams has been a predominantly twentieth-century affair. There were some examples at earlier dates such as the eighteenth-century dams in the Oberharz, while California, a pioneer of numerous dam-building developments because of its particular economic and topographical situation, contains a number of important early rock-fill structures which were built at the end of the nineteenth century.[19]

Early rock-fill dams tended to consist of loose rock-fill cores faced all over with a fitted rubble facing and some kind of watertight upstream coating. Their mode of construction reflected the fact that early in this century labour was cheap and machinery was poor. As this situation changed, and especially as better and bigger machines were developed with which to transport and dump rock, the proportion of loose rock-fill became the predominant constituent, faced with a thin layer of fitted rubble and a watertight upstream membrane. This last was usually of reinforced concrete and sufficiently flexible to allow for settlement, especially at the abutments. By 1931 rock-fill dam-building was an advanced enough technique to allow a height of 328 feet to be achieved in the Salt Springs dam in California.[20] Rock-fill dams have enjoyed a fine safety record and up to 1950 none had failed as a result of any structural defects.

The scale on which twentieth-century dams have been conceived and built, and the vast increase in their numbers, has not only involved a great deal of purely technical activity. The political, economic, social and strategic implications of dams have assumed a major importance in recent decades and have exerted a decisive influence on the course of dam-building history.

Two organisations affording excellent examples of the sort of elaborate relationships which can develop between dam-building and the life of a country are to be found in the United States: the Bureau of Reclamation and the Tennessee Valley Authority.[21]

We have already noted that around 1900 the western states were particularly active in the construction of dams for water supply, irrigation and power. As the scale of these operations grew and the enormous benefits to be derived from them became ever more apparent, it was realised that future progress would need greater investments of capital and more complex forms of co-operation between the states and other organisations involved. The Federal government became the controlling authority in 1902 with the passing of the Reclamation Act and the formation thereby of the Bureau of Reclamation. This new agency was initially concerned with the surveying, planning and construction of irrigation projects, but it soon found itself the controlling authority behind hydroelectric schemes as well. By 1910 the principle was established in the United States that electrical power developed at the Bureau's dams should be sold to cities; the first such case occurred in 1911 when Phoenix, Arizona, began to buy electricity from the Roosevelt dam, the largest masonry dam (it is 284 feet high) ever built up to that date.

Thus in the United States a government department has been responsible for the design, construction and financing of all major dam-building
projects. Only a national organisation of this type could have conceived
and built the magnificent dams which brought the United States to the
forefront of world developments.

The Bureau's first major projects were the Pathfinder and Shoshone
dams, completed before 1910, and then the Roosevelt dam. Between 1912
and 1916 the Bureau's engineers constructed what can now be seen to have
been the prototype of numerous later developments. The Arrowrock
dam on the Boise river in Idaho is a concrete arched dam close to 400 feet
in height and used exclusively for irrigation. Until the Hoover dam
(Pl. 58) was built in 1936 it was the highest in the world. Of the Hoover
dam itself little needs to be said. It is well known as one of the great
monuments of dam-building and its story has been told several times.[22]

In terms of volume of construction, however, the Hoover dam was soon
to be exceeded by the Grand Coulee dam. An interesting point about this
latter is that it was the first structure to exceed in volume the Great
Pyramid of Cheops. It stands to a height of 550 feet across the Columbia
river and creates a reservoir 150 miles long. Like so many twentieth-
century schemes, the Grand Coulee dam is a multi-purpose structure:
it provides water to irrigate 1·2 million acres, it generates 2 million kilo-
watts of electrical power, and it renders the Columbia river navigable for
350 miles up to Revelstoke in Canada. Plate 59 may help to give an impression of this outstanding piece of work.

It is relevant to mention also that the great technical and financial resources of the Bureau of Reclamation allowed them to build some exceptional examples of buttress dams. Structures of this type have been found
to be best suited to special conditions, and in the case of the Stony Gorge
dam (1926–8) in California it was a fault line in the foundations which led
to the construction of a reinforced concrete buttress dam of very slender
and striking proportions. Of the same date is the Coolidge dam in Arizona,
a unique structure comprising three egg-shaped domes supported by
buttresses in the centre and the valley walls at each end. The interesting
thing about the Coolidge dam (Pl. 60) is its extreme use of the cupola
style at such an early date, and yet the idea was not used again in the
United States until recently (e.g. the Glen Canyon dam).

In the year that the Columbia Basin project was begun (1933), President
Roosevelt also brought into existence the Tennessee Valley Authority,
perhaps the best-known example of regional revitalisation and development based on an integrated system of multi-purpose dams.[23] The aim
was to control and harness the Tennessee river in the interests of navigation, flood-control, hydro-power, soil-conservation, fishing and public
health (this last by stamping out malaria). Quite apart from the engineer-

ing work involved in constructing more than thirty dams, the task of the T.V.A. entailed problems of finance, organisation and co-operation on a grand scale, and this is true of the scheme in operation as well as under construction. In order to meet the frequently conflicting requirements of power-generation, flood-control, navigation and malarial extermination with maximum efficiency, the thirty-two dams in the scheme must be operated as a closely integrated unit.

The Tennessee river and its tributaries flow through a large area, and the successful construction and management of the Tennessee Valley project depended on the co-operation of several different states.

The construction and use of dams on certain other large rivers has gone a stage further, and makes demands on the relationships between different countries. The Nile, for instance, is usually thought of in terms of its last 1,500 miles or so, which are in Egypt; it is perhaps easy to forget that a proportion of the main stream, and all of the Nile's headwaters, are in the Sudan and Ethiopia. Thus the construction and use of the Sennar dam on the Blue Nile for the benefit of the Sudanese had to be arranged in conjunction with Egypt, so as to ensure the continued efficient use of the Aswan and Assuit dams. Even more significant was the construction in 1937 of the Gebel Aulia dam on the White Nile, because this in fact was built in the Sudan for Egypt's benefit: it stores extra flood-water to supplement the role of the Aswan dam. The 1929 Nile Waters Agreement was the basis on which the two countries successfully utilised the Nile for twenty-five years, but in 1957 a new agreement had to be negotiated. Plans for the new High Aswan dam threatened to make thousands of Sudanese homeless, while the projected Roseires dam on the Blue Nile threatened to produce a further conflict of water requirements. These plans for new Nile dams resulted eventually in the Sudan being paid £15 million compensation, and it was also agreed that she should have rights to four times as much Nile water as before.

Another of the world's big rivers which has figured in similar negotiations is the Indus. Dam-building on the Indus was difficult even from the technical standpoint (at the Sukkur dam the river can discharge up to 20,000 tons of water per second), and in 1947 the problems were compounded by the creation of the new state of Pakistan. The frontier between India and Pakistan was so arranged that it cut clean through existing dam-based irrigation schemes, and also resulted in the Indus headwaters being in India and the main river in Pakistan. This led to endless disputes about the future use of the Indus water, which were ultimately solved by the intervention of the World Bank, a body which has become intimately involved with the financial arrangements of many big dam projects of recent years.

It was agreed that India should have the unrestricted use of three of the

Indus's eastern feeders while Pakistan should utilise the rest of the basin. The Indus basin is now dammed by some fine structures: the 740-foot-high Bhakra dam, the culmination of an idea which dates from 1906; and the recently completed Mangla dam, 370 feet high and of the earth type. The Mangla dam is shown in Plate 61.

Aluminium owes its commercial availability to electricity; the processes by which the metal is extracted from its ores and subsequently refined require large amounts of electrical power. The first generation of hydro-electric power around 1900 was an early stimulus to the mass production of aluminium, and in turn, as the demand for aluminium's unique properties increased (it is important in aeronautical and electrical engineering), this acted as an important incentive to the development of more hydro-electric projects. The fact that Norway, Canada and the United States became the world's great aluminium producers was in large measure due to these countries being able to generate large quantities of cheap hydro-electric power. Norway is a particularly interesting case; together with countries such as Switzerland, its energy resources are almost 100 per cent hydro-electric. This accounts for the comparatively large number of dams which have been built in Norway and Switzerland and the important developments which have occurred there.

In Great Britain the earliest hydro-electric developments were initiated in 1894 for the production of aluminium.[24] The Foyers project on the east side of Loch Ness was completed in 1896 and is still operated by the British Aluminium Company. Between 1900 and 1910 three more hydro-electric schemes were constructed in North Wales at Ffestiniog, Cwm Dyli and Dolgarrog. The Ffestiniog project was the responsibility of the Yale Electric Power Company and generated electricity for the slate mines and the local town; this was the first time that a public electricity supply had been produced by water power in Great Britain. Neither the Ffestiniog nor the Cwm Dyli project featured dams—they relied on natural supplies from a lake and a waterfall respectively—but for the Dolgarrog project the Aluminium Corporation built two dams.

The first stage of the Eigiau dam was built on the slopes of Carnedd Llewelyn in 1908, and three years later the structure was raised to a maximum height of 35 feet with a crest length of 3,253 feet. It impounded 160 million cubic feet (3,670 acre-feet) of water to drive the Pelton wheels of the Dolgarrog power station and aluminium works. In 1924, in order to provide more storage space, a second dam was built two and a half miles below the Eigiau dam and about a mile to the west of the power station. The Coedty dam was an earth structure made from locally excavated material and equipped with a central core wall of concrete. Its maximum height was 36 feet and it was 860 feet long; the reservoir had a capacity of 11 million cubic feet (252 acre-feet).

On the evening of 2 November 1925 the Eigiau dam failed, apparently as the result of piping. Subsequent investigations[25] showed that it had been badly constructed. The concrete itself was of poor quality, and the foundations of the dam were very inferior. Throughout its length the structure was founded on a deep layer of blue glacial clay, and in places the base of the dam penetrated only a few feet into the top soil and only inches into the clay. At the point where the failure began, the concrete base was only two feet below the top of the clay. Water percolating under the dam is believed to have led to piping, perhaps encouraged by the presence of the low-level outlet pipe and the fact that during the preceding summer, when the reservoir was empty, the clay had partially dried out. Anyway, after seventeen years of trouble-free use a section of clay under the dam was washed away leaving a channel 70 feet wide and 10 feet deep; the dam itself remained intact. The breach was estimated to have been large enough to release 50 million cubic feet (1,150 acre-feet) of water per hour.

The flood swept down the steep valley and poured into the Coedty reservoir. There is nothing to suggest that the Coedty dam was anything but a sound piece of work, but its spillway was designed to deal only with floods arising from rainfall and not with the whole contents of Llyn Eigiau. The Coedty reservoir, being relatively small, was rapidly filled, the spillway overwhelmed and the dam over-topped. The downstream half of the embankment was washed away, the core-wall collapsed and a breach, 200 feet long at the top and 60 feet at the base, released a second wave of water. Without doubt this flood was the more serious because, whereas the Eigiau reservoir emptied at a rate of about 50 million cubic feet (1,150 acre-feet) per hour, the whole contents of the Coedty reservoir, 11 million cubic feet (252 acre-feet), was released at once. The water rushed down the river-bed past the power station and swamped the village of Dolgarrog one mile below in the Conway valley. A great deal of damage was done to the power station and in the village.[26] Fortunately a large number of villagers were attending the weekly film show when the flood wave arrived, and as this was held in the Assembly Hall on high ground, they were unharmed. Otherwise the death toll might well have been more than sixteen. Subsequently the Coedty dam was rebuilt and is still in use, but the Eigiau dam was permanently abandoned.

It is not generally known that another dam belonging to the Aluminium Corporation very nearly came to a disastrous end at about the same time. The Cowlyd dam, built in 1922 in a neighbouring valley, was similar in construction to the Coedty dam. On New Year's Eve 1924 the Cowlyd dam overflowed and a large section of its air face was washed away. Fortunately the reinforced concrete core-wall remained intact and managed to support the water-face half of the embankment. Had the structure failed, the consequences would have been very grave, Llyn Cowlyd being much bigger than either of its neighbours.

The year 1925 was a bad one for British dams. In addition to the troubles in Wales, a small dam in Scotland failed and discharged the $5\frac{1}{4}$ million gallons (19 acre-feet) of water in Skelmorlie reservoir into the Firth of Clyde; several people were killed. The cause of this disaster was curious. The drains of a nearby quarry which emptied into the reservoir had become blocked. When heavy rain filled the quarry with water, the increase in pressure suddenly cleared the choked pipes and flooded the small reservoir below. Its overflow, a small pipe through the embankment, could not cope, and the dam was quickly over-topped and washed away.

Three failures and one near failure in the space of a year had important consequences for British dam-building. There was an immediate spate of public discussion about dams and their potential threat to life and property.[27] Attention was drawn to the recommendations made following the failure of the Dale Dyke dam (see Chapter 9), namely that dams should, by Act of Parliament, be subject to periodic scrutiny. In 1930 the Reservoirs (Safety Provisions) Act was passed, and Great Britain was, it would seem, the first country to take such a step. Essentially the Act requires that all reservoirs exceeding 5 million gallons ($18\frac{1}{2}$ acre-feet) in capacity be inspected at least once every ten years, and that all dams be designed by qualified engineers. The Reservoirs Act took a long time to be fully implemented, however, and as late as 1963 there were still some old dams which had never been examined. Nevertheless defects in a number of dams were brought to light by the Act and steps were taken to make them safe. Dams are the only structures in Great Britain whose security is covered by an Act of Parliament, and since it was passed there have been no major failures. Greatly improved methods of design and construction have, of course, played their part in this enviable record.

Other Welsh hydro-electric schemes of the 1920s were not plagued by failures of the type mentioned above. The Maentwrog scheme of the North Wales Power Company, for instance, begun in 1925 and completed in 1928, features four dams which between them impound Lake Trawsfynnyd. The principal dam of this quartet, built across the Afon Prysor, is of particular interest as it was the first large arch dam ever built in Great Britain. Made of concrete, its geatest height is 96 feet and its base thickness is 37 feet. The dam is shown in Plate 62.

At the end of the First World War the British government set up the Water Power Resources Committee to investigate the direction which should be followed by future hydro-electric developments. The Report clearly showed that the greatest potential lay in Scotland where the Kinlochleven scheme, Britain's first large hydro-electric project, had already been in operation for ten years. A number of dam-based hydro-electric schemes were approved in the early 1920s, but their construction was delayed a decade or more. It was not until the latter years of the 1930s that such schemes as Lochaber, Grampian, Devon and Galloway were fully operational. In most cases the dams were built to raise the levels of natural lochs whose water was then piped to the generating stations below, both of these techniques being especially characteristic of Scottish hydro-electric schemes.

All the above-mentioned Scottish projects were passed by Parliament before 1929; in the next dozen years six more major developments were proposed and all were thrown out. Riparian owners, local authorities and sporting interests vigorously resisted the development of Scotland's hydro-power resources until the beginning of the Second World War. A state of war, however, so threatened imports of fuel, particularly oil, that in 1942 the Cooper Committee was set up to investigate once more Scotland's role in future developments, and it concluded that more projects were feasible and desirable. The North of Scotland Hydro-Electric Board came into being in 1943.

The Board's work has included some of the best British civil engineering-work of recent decades. At the same time scrupulous attention has been paid to conserving the beauty of the Scottish landscape, preserving amenities on various rivers and lochs and making sure that Scotland's important salmon-fishing industry has not been adversely affected. A number of the Board's dams feature elaborate and by no means inexpensive fish passes of various types. From the Board's work in Scotland has stemmed a good deal of basic research into the habits of salmon and the best means to ensure that their migratory life cycle is not disturbed by big dams. The Pitlochry dam, complete with its fish pass, is shown in Plate 63.

The development of hydro-power resources has had a significant effect on the course of dam-building in Great Britain. At the beginning of the twentieth century the traditional water-supply dam, easily the most common type, was either an earth or a masonry gravity structure. Hydro-electric developments, because they were more difficult to justify on economic grounds and because special technical problems were imposed, led to radical departures from earlier practice. The concrete arch dam at Lake Trawsfynnyd was the first of several, notably the ones at Earlston, Tongland and Monar. The dam of Loch Sloy (165 feet high) was the first of a number of concrete buttress dams, one of the largest of the later ones being at the Errochty reservoir. Of the type known as 'diamond headed', its complex stress analysis was checked by photo-elastic methods. Even more revolutionary, however, was the Allt-na-Lairige dam fifty miles north of Glasgow.[28] This structure is 73 feet high, 1,360 feet long and contains only 60 per cent of the amount of concrete which a conventional gravity dam would require. Such economy was achieved by pre-stressing the dam along its water face with groups of iron rods sunk deep into the rock foundations. The Allt-na-Lairige dam was the first pre-stressed dam in the world and, perhaps more than any other structure, typifies the originality, variety and technical interest of Britain's hydro-electric dams.

In the first half of the twentieth century dam-building the world over experienced four major developments: dams became very much bigger in every sense; construction techniques were greatly improved; elaborate methods of analysis led to a more complete understanding of the structural behaviour of dams and better design procedures; and dam engineering in all its phases became subject to a degree of control and organisation which was absent before. It was around 1930 that the full impact of all these influences was properly felt for the first time, and it is significant that since 1930, despite the fact that so many really big dams have been built, the failure rate has dropped sharply. The period between 1920 and 1930 was the last decade in which an appreciable number of dams failed.

Three of these failures were in Great Britain, as we have seen. In Italy the Gleno dam[29] in the Bergamese Alps collapsed on 1 December 1923, the failure subsequently being shown to have been due to high shear stresses and bad workmanship. In Algeria the Habra dam[30] failed on 26 November 1927, a double tragedy in fact because this was the very same structure which had already experienced a collapse in 1881 and in both cases the cause was the same—uplift. On 12 March 1928 the St Francis dam[31] near Los Angeles was completely wiped out when its foundations were washed away. Thus it was that in four major dam-building countries—Great Britain, Italy, France and the United States—official action and legislation came into effect in an effort to ensure that such things could not happen again. These measures, coupled with improved engineering techniques, did indeed carry dam-building into an era of greater reliability. The vast benefits to be derived from dams, whatever their type and however large they might be, are nowadays available at very little risk to life or property.

II

The Last Twenty Years

THIS FINAL CHAPTER can hardly be called an historical one. Nevertheless it is reasonable to bring the story up to date in these last pages and to suggest what elements of recent dam-building work are likely in the future to be judged as being of historical interest and importance.

One fact is beyond dispute. A century ago a few dams were approaching a height of 200 feet, and any over 150 feet were considered exceptional, as Edward Wegmann's remark, quoted at the end of Chapter 9, illustrates. Nowadays the very high dam exceeds 750 feet, while a few are approaching 1,000 feet. In round figures, then, the heights of the highest dams have increased by a factor of 5 within the last century, and the 1960s will go down in history as the decade when the dam of around 1,000 feet in height became a reality. A selection of the world's big dams and large reservoirs is set out in the table overleaf.

Three materials are now established as the basis of dam-building: concrete (for arch, gravity and buttress dams), earth and rock, the last two often in combination. Of the various types, the buttress dam is not in the same class as the others in terms of height, and only one example, the Manicouagan dam in Canada, has yet exceeded 700 feet. It is of the multiple-arch variety.

One exceptional dam deserves to be mentioned. The Nagarjunasagar dam in India is at present being built principally of masonry and, amazingly, by hand. This extraordinary reversion to old-fashioned methods is due to a most singular set of economic conditions, but nevertheless it will ultimately produce a gravity dam 406 feet high and, combined with its end sections of earth, nearly three miles long.[1]

In recent years the stress analysis of concrete dams has been radically

NAME OF DAM	TYPE	HEIGHT (FEET)	LENGTH (FEET)	VOLUME OF DAM (THOUSANDS CUBIC YDS)	RESERVOIR CAPACITY (THOUSANDS ACRE-FEET)	COUNTRY	DATE (U.C.= UNDER CONSTRUCTION)
BHAKRA	Gravity	740	1700	5200	7400	India	1966
BRATSK	Gravity + earth	410	17229	21180	137230	U.S.S.R.	1964
CHIRKEY	Arch	764	1109	1604	2250	U.S.S.R.	U.C.
CONTRA	Arch	754	1246	863	69	Switzerland	1965
FORT PECK	Earth	250	21026	125600	19400	U.S.A.	1940
GARRISON	Earth	210	11300	66500	24500	U.S.A.	1960
GRAND DIXENCE	Gravity	932	2296	7792	324	Switzerland	1962
HIGH ASWAN	Earth + rock	365	11800	57000	127000	U.A.R.	1969
HOOVER	Arched	727	1244	4400	31047	U.S.A.	1936
INGURI	Arch	988	2240	3920	1257	U.S.S.R.	U.C.
KARIBA	Arch	420	2025	1350	130000	Zambia-Rhodesia	1959
KIEV	Earth	62	134180	58030	3024	U.S.S.R.	1964
MANGLA	Earth	370	11050	85000	5880	Pakistan	1968
MANICOUAGAN NO.5	Multi-arch	704	4200	2600	115000	Canada	1967
MAUVOISIN	Arch	780	1706	2655	146	Switzerland	1958
MICA	Earth	800	2550	40000	19800	Canada	U.C.
NAGARJUN-ASAGAR	Gravity + earth	406	15326	73572	9177	India	U.C.
NUREK	Earth + rock	1040	2624	6600	8500	U.S.S.R.	U.C.
OAHE	Earth	246	9250	92000	23600	U.S.A.	1963
OROVILLE	Earth + rock	770	6850	80600	3500	U.S.A.	1967
SAN LUIS	Earth + rock	382		78000		U.S.A.	1967
SAYANSK	Arch	774	3500	11925	12400	U.S.S.R.	U.C.
SOUTH SASKATCHEWAN	Earth	223	16700	86328	8000	Canada	1966
VAJONT	Arch	858	624	460	137	Italy	1961
VOLTA	Rock	370	2100	10350	120000	Ghana	1965

Figure 34 Details of twenty-five dams which are the world's highest, have the greatest volume or form the largest reservoirs.

changed and improved by the application of two new devices: structural models and electronic computers.

Model analysis as a technique for simulating the complex behaviour of a dam goes back to the beginning of the twentieth century. During the 1920s and 1930s model tests of various sorts, sometimes on parts of dams and occasionally on the complete structure, were utilised as a check on theoretical analyses. The use of model dams as a design procedure complete within itself began to be developed in Portugal and Italy around 1950. Portugal's pioneer role in this important new phase in dam design sprang from two factors.[2] The country was firmly committed to developing its hydro-power resources as economically and rapidly as possible, arch dams being selected as the best type in Portugal's mountainous terrain. At the same time there was little native experience of theoretical design procedures such as the trial load method.

Portuguese engineers also subscribed emphatically to the view that only a very full application of the trial-load method would yield an accurate stress analysis. This, they pointed out, was especially so if the structure was unsymmetrical, if it had large gravity abutments, if there were large openings such as might be needed for spillways, or if the foundations were heterogeneous. The engineers of the National Civil Engineering Laboratory in Lisbon advanced the view that in most practical cases the accurate analysis and design of an arch dam by the trial load method entailed a disproportionate expenditure of time, money and effort. They adopted instead the principle that model tests would yield a more faithful simulation of an arch dam's actual behaviour, since all the individual features of any arch dam and its location could be built into a properly fashioned model.

In its early years model testing in various countries—England, Portugal, Italy—saw experiments with a number of different systems. Model dams were made of materials such as plaster of Paris (with various additives), bakelite, plastics such as alkathene, rubber and mortar. The hydrostatic loadings on dams were reproduced with water, mercury or sets of hydraulic jacks. Initially the usual technique of making the model was a casting or moulding process, but in recent years it has become common practice to carve the required shape from a previously cast block. Electric resistance strain gauges, deflection gauges and brittle lacquers are used to measure strains in the model from which the all-important stress distributions in the prototype can be worked out.

At first, model tests were aimed exclusively at the problem of determining stresses due to water pressure. Then, as model analysis became more sophisticated, and as the technique proved to be both valid and advantageous, procedures were developed whereby other problems could be investigated, the four main ones being: the behaviour of variable foundations; the effects of temperature changes; the characteristics of arch dams near and at rupture; and the effects of earthquakes. These new developments in model analysis were all under way by 1960 and were quickly applied to actual design problems, a typical case being the Tang-e Soleyman dam[3] (Pl. 65) in northern Persia, which stands to a height of 330 feet as part of an irrigation scheme.

A particular phase in the evolution of model analysis involved the use of small rubber dams loaded with water, rubber being the only material whose elastic modulus is sufficiently low to allow the use of water as the loading liquid. For various reasons rubber models have now been given up in favour of those made of stiffer materials loaded with either mercury or hydraulic jacks. Nevertheless experiments with rubber models were an integral part of an important piece of research carried out at Imperial College in London in conjunction with the design of the Dokan dam in Iraq. In Chapter 10 it was noted that in 1947 Zienkiewicz had applied relaxation methods to the solution of slices of gravity dams. The work on the Dokan dam, carried out between 1952 and 1955, used the relaxation method to solve the elastic equations of an arch dam: stresses due to water loading and temperature effects were determined.[4] Such a complete stress analysis of a dam of any type had probably not been achieved before, although it must be said that, when one was limited to hand calculation methods, the solution of the large number of equations involved took a very long time.

The Dokan project was the prelude to a rapid and broad extension to other methods of solving the equations of elasticity for dams, gravity and buttress as well as arch. Various shell theories and finite element methods seem to be the approaches attracting most attention and promising the best possibilities for the future. The International Symposium on arch dams held at Southampton University in 1964 indicated the wide range of current ideas and attitudes;[5] and it was most certainly confirmed that any solution of the stresses in a dam by the equations of elasticity involves a great deal of computation, if reliable results, and therefore improved design techniques, are to be obtained. The electronic computer undoubtedly holds the key to future developments in this direction. It alone has the capacity to handle the lengthy and elaborate computations quickly enough to help rather than hinder the job of designing a dam economically. Electronic computers have also markedly improved the usefulness of the long-established trial load method simply by speeding up what was previously a time-consuming procedure.

Research into the behaviour of dams has not been confined to the analysis and design stages. In recent years a number of completed structures have been very fully instrumented so that strains and deflections can be observed in the full-sized article under the full range of operating conditions. A few structures have been to a large extent experiments in themselves. Two of André Coyne's designs, Le Gage dam in France and the Tolla dam in Corsica, were both made exceedingly thin so as to provide a full-scale check on the validity of theoretical analyses when applied to dams approaching the ultimate degree of slenderness. The two dams are a complete success and have been shown to possess adequate margins of safety.[6] Whether or not such thin dams are psychologically justifiable is another matter.

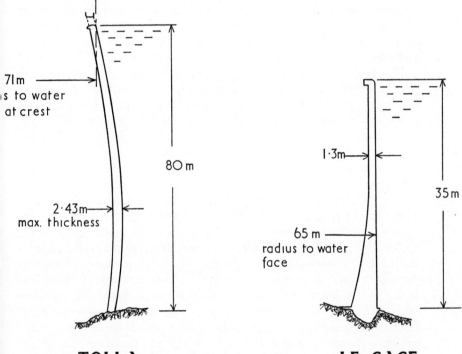

71m
s to water
at crest

80 m

2·43m
max. thickness

1·3m

35m

65 m
radius to water
face

TOLLA   **LE GAGE**

Figure 35   The dams of Tolla and Le Gage, two of the thinnest ever built.

The best-known and most serious dam failure of recent years occurred in 1959. The Malpasset dam, a thin arch structure in southern France, collapsed on the night of 2 December and the small town of Fréjus was swamped. More than 300 people were killed. The subsequent enquiry indicated[7] that the failure had been initiated by weaknesses in the rock on one side of the valley. It was, of course, already well known that arch dams in particular impose very high loadings on their abutments, and attention has been paid to this point for many years. However, a dam failure always results in a re-examination of whatever factor has been the culprit, and the Malpasset accident gave an incentive to the study of rock mechanics. Prior to 1959 only a few books and papers on the subject had been published, but since that year there has been a marked increase in research, discussion and publications. The first International Congress on Rock Mechanics was held in Lisbon in 1966.

The aim of this new field of enquiry is to reduce the design of rock foundations—not only for dams—to as rational and scientific a basis as possible, an objective which is gradually being realised. In the sense that the Malpasset tragedy drew attention, dramatically, to the fact that rock mechanics had validity and application, it can be argued that the failure was not without long-term benefits—something which was true of earlier dam disasters, as we have seen. Parallels can also be found in other branches of civil engineering such as bridge-building.

The development of rock mechanics is symptomatic of an important law of diminishing returns which applies to present and future dam-building. As the years go by, all the best dam sites are rapidly being used up. It is therefore becoming increasingly necessary to attempt the development of sites which are less than perfect and would have been passed over twenty or thirty years ago. Rock mechanics is one field of research whose results are of value in assessing the potential of dam sites which to date have been avoided because their strength is suspect.

Another facet of the same problem is the effect of earthquakes on dams. In the past a very simple solution to the problem of designing against seismic effects was to build in areas where earthquakes do not occur. This is no longer possible. In order to utilise the world's hydro-electric and irrigation resources to the full, it is necessary to construct dams in known earthquake zones, for instance in parts of Persia and India, and in Japan and New Zealand. Like rock mechanics, then, the effects of earthquakes on dams of all types is currently the subject of much research and discussion.[8] Not only is it important to understand the response of each type of dam to seismic activity, but it remains also to determine which sort of dam is best able to resist earthquake effects according to the location and the size of reservoir formed.

There is another aspect of earthquakes and reservoirs which is exceedingly interesting and more than a little disturbing, namely that really large reservoirs can actually *cause* earthquakes. The world's largest man-made lakes such as Mangla, Kariba, Volta and Nasser contain enormous weights of water. Lake Kariba when full weighs something like 170,000 million tons. Such reservoirs, then, impose on the earth's crust prodigious loads where before there was none. In a recent article, Professor J. P. Rothé presented clear evidence[9] that these weights of water can induce earthquakes in areas with no prior history of seismic disturbance.

Examples he quotes include Lake Mead, impounded by the Hoover dam, Lake Kariba which produced two severe tremors in 1962, and the reservoir at Monteynard on the River Drac in France. In 1963 the latter induced a tremor of sufficient magnitude to damage buildings. The most notable and serious case, however, occurred in India in December 1967. The Koyna dam, 340 feet high, impounds 2,800 million cubic metres (2·2 million acre-feet) of water, and as soon as the reservoir began to fill in 1962 seismic activity commenced. It culminated in an earthquake which was measurable all over the world, was damaging throughout India and caused 200 deaths at Koynanagar near Bombay. Evidently the problem is a serious one, and it need hardly be added that if a 'dam-induced' earthquake led to the failure of the dam itself, a double catas-trophe would result.

While the enormous benefits to be derived from modern dams—power generation, irrigation, water supply and river control—are undeniable, a number of other problems, apart from the danger of dam-induced earthquakes, are manifesting themselves, sometimes to a serious degree.

Two points which arose early in this book were the problems of salina-tion and siltation. Both continue to be fundamental difficulties in modern large reservoirs, and no adequate solution has yet been found. Lake Mead was estimated in 1949 to have lost $\frac{1}{37}$ of its capacity to sediment in the first fourteen years of operation. In the case of Lake Mangla, the rate of siltation is estimated to be 1 million acre-feet every twenty years, and this in a reservoir whose ultimate capacity, assuming that plans to heighten the dam at some future date are realised, is less than 9 million acre-feet.[10]

In the case of Lake Nasser loss of storage capacity is only one aspect of the siltation problem. For centuries Egypt's irrigation system utilised the natural fertility of the Nile's silt, and the first Aswan dam was fitted with multiple low-level sluices to ensure the continuation of this technique. But the new High Aswan dam, the structure which forms Lake Nasser, is not so equipped. Thus Lake Nasser will not only act as a gigantic collecting pond for vast amounts of silt but Egypt's agriculture will be adversely affected unless an alternative method of fertilising is introduced. This is possible, of course, but it means spending money on an operation that previously occurred naturally and was free.

While water is standing in a reservoir it is subject to evaporation effects far more serious than would otherwise occur in the river itself. Quite apart from the fact that this water is lost to the atmosphere and cannot therefore be used for power and irrigation, evaporation also leads to an increase in salt concentration. Precisely how serious is the problem of evaporation in large reservoirs is not at the moment very clear, but in the case of Lake Nasser a figure as high as 40 per cent of the total quantity has been quoted as the amount of water likely to be lost. Protagonists of the Lake Nasser scheme suggest that 10 per cent is more reasonable.

Lake Nasser provides the supreme example of the international implications of large dam-building projects and the sort of political repercussions that can result. Initially Egypt was led to believe that either the World Bank, the United States or Great Britain would finance the construction of the High Aswan dam, but in 1956 negotiations broke down. This not only led to the Suez crisis of that year, but it also caused work on the project to be suspended until 1959. Eventually Egypt, unwilling to seek the co-operation of her former advisers and backers, turned to Russia for help. It was with Russian money and technical assistance that the dam was eventually built, but in an atmosphere of sustained criticism and disagreement; there continues to be a body of opinion which believes the project is technically and economically unsound.[11]

The case of Lake Nasser also raises other issues which appear on the debit side of the account, some of which are paralleled in other schemes, and may broadly be classified as social, cultural and strategic.

The diseases bilharzia and malaria are water-borne. Both have been and still are a source of great concern to medical and health authorities, especially in Africa. Large reservoirs are aggravating the problem because the creation of bodies of still water provides new and extensive breeding grounds where there were none before.[12] A good deal is now known about malaria and how to deal with it. In the Tennessee Valley scheme, raising and lowering the water levels of reservoirs during the summer proved an effective control measure. But the same course of action was found to be futile in the case of the Volta lake, and indeed would stimulate the local mosquito to breed even faster. Bilharzia, too, is capable of finding ideal breeding conditions in big reservoirs in Africa. The disease uses as a host small water snails from which it passes to human beings. The transmission is most effectively carried out where there are high densities of snails and humans in static or sluggishly moving water, like the edge of a reservoir. So far the disease has not appeared at either of the reservoirs on the Volta or the Niger. It might, however, especially when the reservoirs fill and spread out to their maximum extent. Bilharzia is also a threat to the success of the High Aswan dam scheme. It is ironical that the very projects designed to improve people's living and working conditions have at the same time aggravated an already serious disease problem.

The establishment of really big reservoirs obviously causes large tracts of land to be flooded. The surface area of Lake Kariba is 1,720 square miles, of the Mangla reservoir 100 square miles, of Lake Volta more than 3,000 square miles and of Lake Nasser nearly 2,000 square miles. Not only is this land lost to agriculture—in those cases, that is, where it was possible to farm in the first place—but other problems arise as well. There is the question of disrupting the life of any towns or villages which lie in the proposed area of the reservoir. In the case of Lake Kariba 50,000 people were involved, 120,000 at Lake Nasser and 70,000 in Ghana at the site of the Volta lake.[13] Of course people can be resettled in other places which in theory offer better prospects for living and working. Yet the habits and customs of a lifetime or even several generations are difficult to change quickly, or if broken completely can be serious for those involved. In deciding the site for a reservoir, then, there are important social issues at stake.

Not only people are deprived of their traditional environment in some big dam schemes. At Kariba (Pl. 64), for example, the threat also involve animals and 'Operation Noah' was staged in order to save wildlife. This exercise became well known all over the world, and it is an encouraging sign that man is concerned to look after animal life in this way, be the specimens rare or not. That this problem could arise again adds yet another dimension to the dam-builder's problem.

In the case of Lake Nasser the overriding cultural problem was the ancient temples and monuments of Nubia. Not everyone is agreed that the Abu Simbel temples were dealt with in the best way, despite the immense cost of moving them and the expenditure of a great deal of international effort to achieve this. But, after a fashion, saved they were for the benefit of posterity and the interest of tourists. Many other ancient buildings and sites were lost, however, and this had been true also in the case of the first Aswan project which flooded the magnificent Temple of Philae when the dam was heightened in 1912. A great deal of emergency archaeological work had to be done in both cases in order to record something more of Egypt's past before the water came to flood, and in the long run to ruin, the relics.

The strategic implications of dams, while not entirely absent in earlier centuries, have assumed a more important role in modern times. The size of reservoirs (and hence their greater destructive potential) and the development of more potent types of aerial warfare are the reasons underlying this situation.

An attempt by the troops of General Franco to blow up the Ordunte dam near Bilbao during the Spanish Civil War was a failure. During the Second World War there were German plans to destroy the Aswan dam, while in 1941 the retreating Russian army blew up the Dnjeprogues dam even though their compatriots were escaping across it at the time.

In March 1943 the Royal Air Force forcibly demonstrated the serious strategic risk to centres of population and industry which the presence of a large reservoir constitutes. The Möhne dam in Germany was badly damaged by one of the most brilliantly conceived raids of the war, although it remains undecided whether or not the long-term effects on German industry were as critical as the British like to believe. The point was made, however; a dam could at least be breached if not actually destroyed by bombs. Quite what this means in an age of nuclear bombs is difficult to assess. It is evident that the destruction of the High Aswan, Kariba or Volta dams would release floods of such stupendous proportions that the results for human life, property, agriculture and industry are too terrible to contemplate. To these would be added the effects of 'radio-active rain' as an unavoidable by-product. Indeed, it has been pointed out that a nuclear device exploded in a large reservoir would achieve such widespread contamination through 'rain' that attacks on dams themselves are no longer the issue.

Modern large-scale dam-building operations, then, have their draw-backs, problems and defects. It is natural to wonder, therefore, to what extent the use of dams may in the future alter or even diminish. Is it conceivable that dams will sooner or later cease to be built at all? [14]

Man's dependence on dams as the means of harnessing and controlling rivers is not likely to diminish in the near future.[15] The benefits to be derived, especially in underdeveloped countries, are so obvious and worth-while that the various political, financial and social problems will be overcome if humanly possible, and the strategic risks will either be lived with or ignored. There will be technical problems, too. They may well be more difficult to solve than those of the past, but solved they will be; that is in the very nature of engineering.

At the same time, however, it must be recognised that viable dam sites will ultimately run out. Countries such as Switzerland and Norway have already harnessed a very high proportion of their potential hydro-electric resources, and it has been suggested that further developments must soon be based on some other fuel, probably nuclear. In other countries also, nuclear power may soon offer an economic alternative to hydro-electric development. Nevertheless something below 5 per cent of the world's hydro-power resources have so far been realised and, of the remainder, dams will surely be called upon to yield a few more per cent at least.

Of all the water on the earth's surface a mere 2 per cent is fresh, and most of that is held in the ice caps. To meet further irrigation and water-supply needs, dams will continue to be used until such time as the purifi-cation of sea-water becomes an economic proposition; in all but a few places, where special conditions prevail, this is still some way off.

There seems little doubt, then, that in the future dams will continue to contribute to man's efforts to control, harness and manipulate his environment. Whatever course the development of dams may take, they will be entitled to claim a long and splendid ancestry. It is a heritage which deserves to be fully studied and recorded, and the author offers this book as a first contribution.

Notes

(Some of the bibliographical references are complete within the following Notes. For the remainder, please consult the Bibliography.)

Chapter 1 Antiquity

1. MURRAY, p. 172. Murray's study of the dam is the best one in English.
2. See, for instance, HELLSTRÖM and HATHAWAY (1)
3. It was G. Schweinfurth, the discoverer of the dam, who linked its use with the alabaster quarries.
4. For information on ancient quarrying, see FORBES (1), Vol. VII, pp. 162–77
5. For an account of basin irrigation, see FORBES (1), Vol. II, pp. 2–5, 22–30
6. HELLSTRÖM shows eleven photographs of the dam's remains.
7. A recent claim, due to Professor J. Vercouter, that there was also a dam across the Nile at Semna is discussed by KEATING, p. 37. For the improbability of this, see GOBLOT (1), p. 48
8. The edition of Herodotus used throughout is HERODOTUS, *The History of Herodotus*, trans. by G. Rawlinson, 2 vols, London 1910
9. Among those who have fallen for the story are PRELINI, p. 556 and SCHNITTER, p. 142
10. See G. CATON-THOMPSON and E. W. GARDNER: 'Recent Work on the Problem of Lake Moeris' in *The Geographical Journal*, Vol. 73, 1928, pp. 20–58
11. Studies of the rivers are in IONIDES and GEOGRAPHICAL HANDBOOK SERIES, No. 524, Chapters II and X. For a map of the delta, see Fig. 8
12. For an account of perennial irrigation, see FORBES (1), Vol. II, pp. 2–5, 16–19
13. L. DELAPORTE, *Mesopotamia. The Babylonian and Assyrian Civilisations*, London 1925, pp. 105–7
14. WILLCOCKS (1), p. 140
15. See FORBES (1), Vol. I, pp. 155–9 for a description of Sennacherib's dams.
16. THOMPSON and HUTCHINSON, pp. 114–16
17. STRABO, XVI, 1, 9 (Loeb Edn, Vol. VII, p. 205, quoted by courtesy of William Heinemann Ltd)
18. For details of the Sabaeans, see P. K. HITTI, *History of the Arabs*, London 1967, Chapter V

19. E. GLASER, *Reise Nach Marib*, Vienna 1913. A useful résumé is in the ENCYCLOPAEDIA OF ISLAM, Vol. III, p. 286

20. Plan of the dam site based on Dr Fakhry's work is in G. KHEIRALLAH, *Arabia Reborn*, Albuquerque 1952, p. 21

21. BOWEN and ALBRIGHT, pp. 70–80

22. For comments on the references in Pliny and the Koran, see D. L. O'LEARY, *Arabia before Mohammed*, London 1927, pp. 89–92

23. See GOBLOT (2), p. 110

24. Described by VINCK, p. 356

25. ENCYCLOPAEDIA OF ISLAM, Vol. III, p. 289

26. For the Himyarite and Jewish dams, see TWITCHELL, pp. 68, 116

27. The main source of information is KEDAR

28. GLUECK (1), pp. 160–6, includes two photographs and a plan.

29. An archaeologist friend of the author told him that when this dam failed a few years ago several people were drowned.

30. See Bibliography

31. See, for instance, FORBES (2), pp. 19–22

32. PRELINI believed that they did, but his arguments are highly speculative and based on false information.

33. It must be remembered, though, that a dam failure could cause crop failure, and in this sense a stable structure was important.

Chapter 2   The Romans

1. For 'pozzolana', see Glossary

2. See PARKER, pp. 57–60

3. GIOVANNONI, pp. 275–8

4. To be precise, it was the approach spans which Giovannoni examined. The main spans fell with the dam when it collapsed.

5. See Sir T. ASHBY, pp. 253–6

6. MARION E. BLAKE, *Roman Construction in Italy*, Carnegie Institution Publication No. 616, Washington 1959, pp. 85–6

7. SEXTUS JULIUS FRONTINUS, Book II, Section 93. See FRONTINUS, *The Stratagems and Aqueducts of Rome*, Loeb Classical Library, London 1961, p. 421 (quoted by courtesy of William Heinemann Ltd)

8. This is the view of O. BONAMORE, *Il Sacro Speco e S. Scolastica*, Venezia 1884, p. 159

9. R. LANCIANI, *Ancient Rome in the Light of Recent Discoveries*, Boston 1889, p. 274

10. The work of Calvet and Nicolas is discussed by BENOIT, p. 339, and GOBLOT (2), pp. 114–18

11. This drawing is reproduced opposite p. 333 of BENOIT's article.

12. Henri Goblot has tried hard to track down this material and also that of Calvet and Nicolas but sadly without success; see GOBLOT (2), pp. 113–20

13. SALADIN, pp. 162–4

14. This is the view of GOBLOT (2), p. 112

15. See GOBLOT (2), p. 112

16. The main source of information is VITA-FINZI

17. C. S. JARVIS, *Desert and Delta*, London 1938, Chapter XI

18. Sir A. STEIN: 'Surveys on the Roman Frontier in 'Iraq and Trans-Jordan' in *The Geographical Journal*, Vol. XCV, No. 6, June 1940, p. 430

19. STRABO, XVI, 2, 19 (Loeb Edn, Vol. VII, p. 263, quoted by courtesy of William Heinemann Ltd)

20. See DUSSAUD, p. 137

21. CONDER (1), pp. 171 *et seq*

22. BROSSÉ's article contains numerous drawings and photographs.

23. These facts are due to Talmudic writers. See, for instance, CONDER (2), p. 255, and A. NEUBAUER, *La Géographie du Talmud*, Paris 1868, pp. 29, 299

24. Of course the dam was not built with a tapered crest. Over the centuries the earth cover has been eroded and never fully replaced.

25. CASADO, p. 360

26. The most complete set of dimensions are given by CASTRO GIL

27. This is the view of F. J. WISEMAN, *Roman Spain*, London 1956, p. 163

28. CELESTINO GÓMEZ, p. 561

29. CASADO, pp. 361–3

30. CASADO, p. 358

31. R. THOUVENOT, *Essai sur la Province Romaine de Bétique*, Paris 1940, pp. 522–5

32. MOORE, pp. 98, 102

Chapter 3 Byzantium and Persia

1. GOBLOT (1), p. 51

2. See the article by SMITH

3. PROCOPIUS, *The Buildings*, II, iii, 16–21 (quoted by courtesy of William Heinemann Ltd)

4. PROCOPIUS, *The Buildings*, II, x, 17 (quoted by courtesy of William Heinemann Ltd)

5. GLUECK (2), pp. 80, 209

6. HOUTUM-SCHINDLER, p. 288

7. They are mentioned by GOBLOT (1), p. 48

8. Sir A. T. WILSON, *The Persian Gulf*, London 1954, p. 35

9. A good account of the dams at Shushtar is in CURZON (1), pp. 706–13

10. For a picture of this dam, see L. M. WINSOR: 'Irrigation in Iran' in *Civil Engineering*, Vol. 14, February 1944, p. 63

11. They are mentioned by CURZON (1), p. 715, and there is a drawing of one of them in LYNCH, p. 594

12. CURZON (1), pp. 492–3

13. Taken from KLEMM, p. 79

14. Sir A. STEIN: 'An Archaeological Journey to Western Iran' in *The Geographical Journal*, October 1938, p. 327

15. For a general discussion, see ADAMS (1), pp. 116–19

16. A considerable contribution, however, is that of ADAMS (2)

17. WILLCOCKS (2), p. 10. Included are details of the canal.

18. See ADAMS (2), Plates 18, 19

19. For more details, see JACOBSEN and ADAMS, p. 1257

20. A. K. S. LAMBTON, *Landlord and Peasant in Persia*, Oxford 1953, p. 91

21. LE STRANGE (1), p. 213

22. GOBLOT (1), p. 49, and (3), pp. 43, 49

23. JUSTIN and TALEGHANI, p. 157

24. GOBLOT (3)

25. It is interesting that a law of increasing returns operates when a dam is heightened. At Kebar an increase in height of 13 per cent gave an increase in storage capacity of perhaps as much as 40 per cent.

26. GOBLOT (2), p. 129. In 1960 M. Goblot had been unable to establish anything definite.

27. WULFF, p. 246

28. CURZON (2), Vol. 2, p. 316

29. Dimensions given by WULFF, p. 248

30. JUSTIN and TALEGHANI, p. 154

31. For a description of this and other Persian bridge-dams, see H. SHIRLEY-SMITH, *The World's Great Bridges*, London 1953, pp. 25–8

Chapter 4 The Moslems

1. There are certain well-known exceptions such as the Moslems' use of the windmill and the fact that they brought the arts of paper-making and gunpowder manufacture to Europe.

2. TWITCHELL, pp. 23, 65

3. LE STRANGE (2) is based on Ibn Serapion's work and also that of other Moslem geographers.

4. The dams at Kubbin are mentioned in LE STRANGE (2), pp. 44, 46

5. LE STRANGE (1), p. 58. This volume is a mine of information on Moslem technology, and so too are Le Strange's translations of Moslem writers such as Ibn Serapion, Ibn-al-Balkhi, Mustawfi and Mukaddasi.

6. See also WILLCOCKS (2), Plate 7, and for further information, ADAMS (2), pp. 77–8

7. It is illustrated in E. F. SCHMIDT, *Flights over ancient cities of Iran*, Chicago 1940, Plate 56 and p. 44

8. LE STRANGE (1), p. 239

9. This extract is from KLEMM, p. 79

10. IBN-AL-BALKHI, p. 876. His *Fars Namah* (Book of Fars) was compiled about A.D. 1107.

11. IBN-AL-BALKHI, p. 869

12. The full extract is in WULFF, p. 246

13. IBN-AL-BALKHI, p. 28

14. These and other aspects of irrigation on the Kur are discussed by HOUTUM-SCHINDLER

15. IBN-AL-BALKHI, p. 869

16. All the information is from LE STRANGE (1)

17. For Moslem surveying techniques, see E. R. KIELY, *Surveying Instru-*

ments, their history and classroom use, New York 1947, Chapter III, and L. OLSON and H. L. EDDY: 'Ibn-al-Awam, a soil scientist of Moorish Spain' in *The Geographical Review,* 1943, p. 107

18. See, for instance, the lists prepared by IMAMUDDIN, pp. 127–31. Imamuddin's book is a splendid background study for historians of technology.

19. IMAMUDDIN, p. 104

20. AYMARD, Chapter I, and MARKHAM, Chapter X, are the basic sources of information.

21. AYMARD, p. 20

22. For details of the dam and irrigation scheme, see PASSA, pp. 127–37 and Plate II

23. SCOTT-MONCRIEFF, pp. 142–51

24. MARKHAM, Chapter VIII. Although not dated, the book was written in 1867. It is not clear when Markham was in Spain.

25. MARKHAM, Chapter VIII

26. AYMARD, Chapter XV is the principal source of information to which the author has added his own conclusions based on a visit in 1966.

27. Orihuela's irrigation is described by MARKHAM, Chapter IV

28. CAVANILLES, Vol. 2, p. 264

29. PRELINI, p. 557, and MARKHAM, p. 46, are examples.

30. SCOTT-MONCRIEFF, p. 124

31. LLAURADO, p. 489

32. The names 'Cornalvo' and 'Alcantarilla' continue to be used.

Chapter 5 Christian Spain

1. For information on them, see W. MONTGOMERY WATT, *A History of Islamic Spain,* Edinburgh 1965, pp. 150–4, 162–4

2. SCOTT-MONCRIEFF, p. 171

3. ALZOLA Y MINONDO, p. 92

4. BRUNHES, p. 125

5. LLAURADO, p. 686

6. GOBLOT (2), pp. 124, 126

7. AYMARD, Chapter VII. The essence of Aymard's information on Spanish dams is in WEGMANN, Chapter VII

8. AYMARD, p. 122

9. Such as ALZOLA Y MINONDO, p. 93, SCHNITTER, p. 145, GOBLOT (2), p. 124. PRELINI goes so far as to claim the dam for the Moslems.

10. AYMARD, Plate V. Monsieur Aymard's set of drawings, beautifully drawn, are in a separate volume from the text.

11. For an account, unsigned, of the dam's early history, see REVISTA DE OBRAS PUBLICAS (1), pp. 342–5, 350–1

12. For information on how the project was financed, see ESTRADA, p. 14, and ARGUELLES, p. 377

13. It is worth quoting the height in *palmos* since a figure of 200 *palmos* indicates an arbitrarily chosen height, whereas $137\frac{1}{4}$ feet would suggest something very precise.

I

14. Sets of drawings of the dam will be found in AYMARD, Plates VI and VII, and LLAURADO, pp. 167–74

15. REVISTA DE OBRAS PUBLICAS (1), p. 344, and CAVANILLES, Vol. 2, p. 185

16. IBARRA Y RUIZ, p. 80. This is a basic work on the history of Elche's irrigation.

17. CAVANILLES, Vol. 2, p. 274

18. ALZOLA Y MINONDO, p. 238

19. It is surprising that SCHNITTER, p. 145, regards the Relleu dam as having a 'bizarre shape'. In fact it has a simple shape, entirely appropriate to the site.

20. The capacities are: Proserpina, 2,800 acre-feet; Alicante, 3,000 acre-feet approx. See ICOLD, Vol. 1, Spain, p. 1, for full statistics of old Spanish dams.

21. REVISTA DE OBRAS PUBLICAS (1), p. 342. Another mention is in GLICK, p. 166

22. The Codex is entitled *Los veinte y un libros de los yngenios e maquinas de Juanelo*. See L. RETI.

23. For example LLAURADO, p. 167 and ALZOLA Y MINONDO, p. 142

24. IBARRA Y RUIZ, pp. 24, 80

25. LLAURADO, p. 346, and BRUNHES, p. 109

26. CAVANILLES, Vol. 2, p. 259, and COMISION INTERNACIONAL DE RIEGOS Y DRENAJES

27. See ARGUELLES, p. 378, and BECCERIL, pp. 389–90. In 1926 the dam was raised to 27·3 metres and is now called Arguis.

28. The section on dams has been published by AGUILA (1)

29. AGUILA (1), p. 191

30. AGUILA (1), p. 194

31. See the account of the dam by LAZARO URRA, p. 218

32. The principal source is AYMARD, Chapter XIX

33. See MUSSO Y FONTES. The account is reproduced in AYMARD, p. 257, and a shortened version is given by WEGMANN, p. 58. A somewhat fanciful local tale based on the accident appears in MARKHAM, p. 25

34. ALZOLA Y MINONDO, p. 347

35. REVISTA DE OBRAS PUBLICAS (2), p. 155

36. AGUILA (2), p. 6

Chapter 6 The Early Americas

1. See M. DAUMAS, *Histoire Générale des Techniques*, Vol. I, Paris 1962, pp. 391–423

2. KOSOK, p. 174. Dr Kosok's article is a general study of early Peruvian agriculture.

3. SQUIER, p. 218

4. *Ibid.*

5. STEPHENS, Vol. 2, Chapters XII and XIII, and p. 249. See also WITTFOGEL, pp. 184–8, for Maya irrigation.

6. V. W. VON HAGEN, *World of the Maya*, New York 1960, pp. 25, 63. More details about this dam would be welcome.

7. A good account of the technique is M. D. COE: 'The Chinampas of Mexico' in *Scientific American*, Vol. 211, July 1964, pp. 90–8

8. RUDOLPH, p. 531. This article is a basic source of information on Potosí and its dams.

9. *Ibid.*

10. RUDOLPH, p. 537

11. GOMEZ-PEREZ, p. 27

12. H. G. WARD, *Mexico in 1827*, London 1828, Vol. 2, pp. 282–8. The structure still exists and now carries a road; see G. C. VAILLANT, *The Aztecs of Mexico*, 1950, p. 262

13. GOMEZ-PEREZ, p. 25

14. *Ibid.*

15. See the article by HINDS (1). Another dam, that of Saucillo, is illustrated in HINDS (2), Fig. 5

16. HINDS (1), p. 251

17. A photograph in GOMEZ-PEREZ, p. 25, shows this condition.

18. HINDS (1), p. 252. What appears to be the same dam is illustrated by GOMEZ-PEREZ, p. 26

19. HINDS (1), p. 253

20. WOODBURY, pp. 62–5, and References 9, 10, p. 84

21. HARRINGTON, p. 380, has some information on works around San Antonio.

22. SCHUYLER, p. 125. There is a picture in G. D. CLYDE: 'Irrigation in the United States' in *Transactions*, ASCE, Vol. CT, 1953, p. 316

23. See UHL, p. 147, and HATHAWAY (2), pp. 478–9

24. WEGMANN, Part II, Chapter 4, discusses the construction of timber dams in the United States.

25. HATHAWAY (2), p. 478

Chapter 7 Europe

1. SKEMPTON (1), pp. 444–5

2. A good deal of information on early dams around Milan will be found in BAIRD-SMITH, Vol. 1

3. BAIRD-SMITH, Vol. 1, p. 64, and Plate IV

4. PARSONS, p. 327

5. PARSONS, p. 352, and Fig. 128

6. The source from which some of the information on the Cento and Ternavasso dams has been taken is classified. The author was allowed to study the historical sections with the permission of the Istituto Idraulico of the Politecnico di Milano on the understanding that no reference was given.

7. See note 6.

8. There are several sources, notably NOETZLI, LANSER, pp. 153–5, and C. SEMENZA: 'Dighe ad arco e cupola' in *Wasser-und Energiewirtschaft*, 1956, p. 213

9. NOETZLI, p. 452
10. LANSER, pp. 153–6
11. LANSER, p. 150
12. AGRICOLA, *De Re Metallica*, translated by H. C. and L. H. HOOVER, New York 1912
13. ZIEGLER, pp. 125–6
14. An example is shown in FERRENDIER, Part IV, p. 783. On pp. 771–4 is more useful information on old French mill-dams.
15. L. TALABOT: 'Le Moulin de Bazacle' in *Mémoires de l'Académie des Sciences*, Toulouse 1863, pp. 326–33, plus figures.
16. Such as those in KLEMM, Fig. 12; J. K. FINCH, *Engineering News Record*, 1930, p. 680; and R. S. KIRBY and P. G. LAURSON, *Early Years of Modern Civil Engineering*, New Haven 1932, p. 191
17. Sources such as SKEMPTON (1), p. 468, and ANDREOSSY, pp. 104–14, have been used.
18. FROIDOUR, pp. 16 *et seq*. Froidour's description is the earliest of a dam being built.
19. ANDREOSSY, pp. 114–18, is a useful account of the dam written soon after its completion.
20. See the article by HILARY M. PEEL
21. Useful surveys are WILSON (1) and (2).
22. A basic study is that of WILLAN
23. A very thorough account is by W. B. STEPHENS
24. A history of the firm is R. H. CAMPBELL, *Carron Company*, Edinburgh 1961. The water-supply problem is discussed on pp. 42–5
25. SMEATON, Vol. 1, pp. 252–3
26. SMEATON, Vol. 2, pp. 63–8
27. SMEATON, Vol. 2, p. 65

Chapter 8 The First Half of the Nineteenth Century

1. Basic data on early British dams will be found in ICOLD, Vol. 1, Great Britain, p. 1.
2. For discussion of this point, see WALTERS (1), pp. 55–8
3. The dam is mentioned in HADFIELD (1), pp. 240–1, 327
4. Compensation water is discussed by KENNARD, p. 495, and WALTERS (2), pp. 107–18, 202–3
5. For Telford's work on canal dams, see L. T. C. ROLT, *Thomas Telford*, London 1958, pp. 166–8, 178–9, 183 n
6. See HADFIELD (2), pp. 88–90
7. There is an account with photographs in WALTERS (1), pp. 42–5
8. Quoted in HADFIELD (1), p. 230 (by permission of David and Charles, Newton Abbot)
9. A very full account with many diagrams is by GUENOT
10. SKEMPTON (2), pp. 91–5, and Plates IX-XII
11. The book by O'CONNOR is an account of the disaster. See also CONDIT (1), pp. 258–9

12. See the article by LEGGET. The same author also wrote *Rideau Waterway*, Toronto 1955

13. An entertaining general account is F. W. ROBINS, *The Story of Water Supply*, London 1946

14. Interesting information is in R. A. PAXTON, *Three Letters from Thos. Telford*, Edinburgh 1968. See also WALTERS (2), pp. 93–4

15. The background is given in WALTERS (2), pp. 96–110

16. KENNARD, p. 496

17. See GOBLOT (2), pp. 137–9, and WEGMANN, p. 64. A cross-section and plan are shown in COYNE (1), p. 203

18. A full account of the scheme is by NELSON M. BLAKE, *Water for the Cities*, New York 1956. This is a basic study of the history of water supply in the U.S.A.

19. See WEGMANN, pp. 153–4

20. WILSON (2), p. 311. There is also information on nineteenth-century water-power dams.

21. See RAISTRICK and JENNINGS, Chapter 9

22. ROUSE and INCE, pp. 165–6

23. WEGMANN, pp. 290–2, and Plate XCVII (*not* CXVII)

24. The broad picture is presented in ADDISON, Chapters 5–10, and NORRIE, Chapter VIII

25. The Nile delta works are covered in WEGMANN, pp. 101–4, and A. G. VAUGHAN-LEE: 'The Mohammad Aly Barrages, Egypt' in *Jour. Inst. Civ. Engrs.*, Vol. 15, 1940–1, pp. 237–79

26. The Coleroon dams are described by NORRIE, pp. 141–2, and in CENTRAL BOARD OF IRRIGATION AND POWER, pp. 61–3

27. G. B. WILLIAMS, *Storage Reservoirs*, London 1937, p. 142. Figure 76 is based on Williams and WEGMANN, p. 124

Chapter 9   The Second Half of the Nineteenth Century

1. SIMON STEVIN, *De Beghinselen des Waterwichts*, 1586; see *The Principal Works of Simon Stevin*, edited by E. J. Dijkterhuis, Amsterdam 1955, Vol. 1, pp. 421–73

2. SIMON STEVIN, *Nieuwe Maniere van Sterctebou Door Spilsluysen*, 1617; see *The Principal Works of Simon Stevin*, Vol. 5, Engineering, edited by R. J. Forbes, Amsterdam 1966, pp. 113–21

3. B. F. DE BELIDOR, *Architecture Hydraulique*, Vol. 1, Part II, Paris 1750, pp. 80–3

4. They are quoted in AGUILA (1), pp. 188–94

5. One of Bossut's drawings can be found in FERRENDIER, Part IV, p. 774

6. For DE SAZILLY's paper, see Bibliography.

7. The project is described by WEGMANN, pp. 65–9

8. For DELOCRE's paper, see Bibliography.

9. GRAEFF, pp. 198, 207–11

10. GRAEFF, Fig. 10

11. GRAEFF, p. 196

12. For RANKINE's paper, see Bibliography.

13. For a selection, see BELLET, 2nd part, and LELIAVSKY (1), pp. 500–2

14. WEGMANN, Chapter III

15. WEGMANN, pp. 94–8, BELLET, pp. 54–6, and LE GÉNIE CIVIL (see Bibliography).

16. For details, see BELLET, pp. 65–70, and SCHUYLER, pp. 258–9

17. See LELIAVKSY (2), Part I, pp. 1–11

18. G. F. DEACON: 'The Vyrnwy Works for the Water-Supply of Liverpool' in *Min. Proc. Instn. Civ. Engrs.*, Vol. CXXVI, 1895–6, Part IV, p. 28

19. DELOCRE, Chapter II, pp. 248–59

20. ZIEGLER, pp. 294–5

21. SCHUYLER, pp. 163–74, and J. M. ALARCO: 'The Bear Valley Dam as an Arch' in *The Technograph*, University of Illinois, No. 14, 1899–1900, pp. 105–10

22. See SCHUYLER, pp. 126–52

23. Descriptions of all these dams are in SCHUYLER, Chapter III

24. L. A. B. WADE: 'Concrete and Masonry Dam-Construction in New South Wales' in *Min. Proc. Instn. Civ. Engrs.*, Vol. CLXXVIII, 1908–9, Part IV, pp. 1–110

25. A good example is the papers and discussion on dam design in the *Minutes of Proceedings*, Vol. CXV, 1893–4, Part I, pp. 12–188

26. Its construction is described by SCHUYLER, pp. 189–205

27. Typical lists are in JUSTIN, pp. 3–6, and T. A. MIDDLEBROOKS: 'Earth-dam practice in the United States' in *Transactions*, ASCE, Vol. CT, 1953, pp. 700–4

28. The report is entitled *Report on the Failure and Bursting of a RESERVOIR EMBANKMENT belonging to the SHEFFIELD WATERWORKS COMPANY, on the Night of Friday, 11 March 1864*. It was prepared by 'Robert Rawlinson and Nathaniel Beardmore, Civil Engineers'.

29. See, for instance, Sir John James HARWOOD, *History and Description of the Thirlmere Water Scheme*, Manchester 1895

30. See ROUSE and INCE, Chapters X and XI; also F. W. KEATOR: 'Benoit Fourneyron (1802–1867)' in *Mechanical Engineering*, April 1939, p. 295

31. The picture is set out in P. DUNSHEATH, *A History of Electrical Engineering*, London 1962

32. See M. KRANZBERG and C. W. PURSELL, JR., *Technology in Western Civilisation*, Vol. 1, Oxford 1967, pp. 582–92

33. A good account of the Austin dam is SCHUYLER, pp. 242–51

34. For information on early Italian dams, see MINISTERO DEI LAVORI PUBBLICI, pp. XXX–XXXV

35. WEGMANN, p. 40

Chapter 10   The Twentieth Century

1. See L. W. ATCHERLEY and KARL PEARSON, *On Some Disregarded Points in the Stability of Masonry Dams*, University of London, Drapers' Company Research Memoirs, Technical Series II, 1904

2. See PIPPARD, pp. 265–77

3. See J. S. WILSON and W. GORE: 'Stresses in Dams: an Experimental Investigation by means of India-Rubber Models' in *Min. Proc. Instn. Civ. Engrs.*, Vol. CLXXII, 1907–8, Part II, pp. 107–33

4. Useful background information on elastic solutions is in LELIAVSKY (1), pp. 502–23

5. O. C. ZIENKIEWICZ: 'The Stress-Distribution in Gravity Dams' in *Jour. Instn. Civ. Engrs.*, Vol. 27, 1946–7, pp. 244–71

6. The project is described by M. FITZMAURICE: 'The Nile Reservoir, Assuan' in *Mins. Proc. Instn. Civ. Engrs.*, Vol. CLII, 1902–3, Part II, pp. 71–155

7. See WILLCOCKS (3)

8. Some thoughts on this point are in LELIAVSKY (1), Chapter 12, Section XI

9. Full details are set out in U.S. BUREAU OF RECLAMATION, *Treatise on Dams*, Denver 1950, Chapter 10

10. COYNE (2)

11. See C. SEMENZA: 'Arch Dams: Development in Italy' in *Journal of the Power Division*, ASCE, Vol. 82, 1956, Paper 1017; also COYNE (3), p. 9597

12. See E. W. LANE: 'Security from Under-Seepage Masonry Dams on Earth Foundations' in *Transactions*, ASCE, Vol. 100, 1935, p. 1237

13. A. C. KOENIG: 'Dams on Sand Foundations' in *Transactions*, ASCE, Vol. LXXIII, 1911, pp. 175–224

14. JUSTIN, Chapters V, VI. Justin's book is typical of early texts on earth-dam design.

15. See BOSTON SOCIETY OF CIVIL ENGINEERS, pp. 295–336, for Cassagrande's paper.

16. See TERZAGHI, pp. 679–82 (see also pp. 687–90 for history of piping)

17. The point is mentioned in J. L. SHERARD *et al.*, *Earth and Earth-Rock Dams*, New York 1963, p. 325. Other historical aspects of stability analyses are discussed.

18. This technique is dealt with by SCHUYLER, Chapter II

19. See SCHUYLER, Chapter I

20. The Salt Springs dam is covered in C. V. DAVIS, *Handbook of Applied Hydraulics*, New York 1942, Section 8, pp. 289–332

21. The USBR and the TVA are described by CONDIT (2), pp. 231–73

22. In, for instance, *The Story of Hoover Dam*, Ingersoll-Rand Company, New York 1932–6

23. A history of the scheme in all its aspects is D. E. LILIENTHAL, *TVA: Democracy on the March*, New York 1944

24. A valuable account of British hydro-electric developments is J. GUTHRIE BROWN: 'Sixty years of Hydro-electric Development in Great Britain' in *The Structural Engineer*, November 1956, pp. 373–403

25. See J. GUTHRIE BROWN: 'Discussion on Dam Disasters' in *Proc. Instn. Civ. Engrs.*, Vol 27, 1964, pp. 366–8; also *Engineering*, Vol. 120, 1925, pp. 581, 711, 810

26. There is an eye-witness account in G. GERARD, *Hydro-Electric Engineering*, London 1949, pp. 150–1

27. See 'Discussion on Dam Disasters' in *Proc. Instn. Civ. Engrs.*, Vol. 27, 1964, pp. 343–76, for comments on the effects of the dam failures of 1925.

28. The dam is described in J. A. BANKS: 'Allt-na-Lairige Prestressed Concrete Dam' in *Proc. Instn. Civ. Engrs.*, Vol. 6, 1957, pp. 409–44

29. The Gleno dam is discussed in *Engineering News-Record*, 31 January 1924, p. 182

30. For the Habra dam's failures, see the reference to LE GÉNIE CIVIL in the Bibliography.

31. The full story of the St Francis dam disaster is in C. F. OUTLAND

Chapter 11 The Last Twenty Years

1. See *Engineering News-Record*, 8 June 1967, pp. 44–5

2. For Portuguese attitudes to arch-dam design, see M. ROCHA *et al.*: 'Arch Dams: Design and Observation of Arch Dams in Portugal' in *Journal of the Power Division*, ASCE, Vol. 82, 1956, Paper 997

3. R. G. T. LANE and J. L. SERAFIM: 'The Structural Design of Tang-e Soleyman Dam' in *Proc. Instn. Civ. Engrs.*, Vol. 22, July 1962, pp. 257–89

4. D. N. DE G. ALLEN *et al.*: 'The Experimental and Mathematical Analysis of Arch Dams' in *Proc. Instn. Civ. Engrs.*, Part I, Vol. 5, May 1956, No. 3, pp. 198–258

5. See J. R. RYDZEWSKI, *Theory of Arch Dams*, London 1965, for the collected papers of the Symposium.

6. They are described in COYNE (4), pp. 275–8

7. For an expert's summary of the official enquiry, see C. JAEGER: 'The Malpasset Report' in *Water Power*, Vol. 15, 1963, pp. 55–61

8. See, for instance, *Transactions*, Ninth International Congress on Large Dams, Istanbul, 1967, Vol. IV, Question No. 35

9. J. P. ROTHÉ: 'Fill a lake, start an earthquake' in *New Scientist*, Vol. 39, No. 605, 11 July 1968, pp. 75–8

10. For a detailed account of the Mangla project, see *Proc. Instn. Civ. Engrs.*, Vol. 38, November 1967, pp. 345–576. Siltation is mentioned on pp. 349–50

11. On this point, see the article 'The Aswan High Dam' in *Water Power*, August 1965, pp. 301–9, and October 1965, pp. 385–6

12. See, for instance, LOWE-MACONNELL, pp. 87–94. Many other problems of modern large reservoirs are discussed in this comprehensive book.

13. Aspects of resettlement are discussed in WARREN and RUBIN, Chapter III. This book is a valuable social, economic and legal study of Africa's large reservoirs.

14. This notion has been advanced more than once, e.g. by L. DUDLEY STAMP: 'Dams and deserts—are our concepts wrong?' in *Jour. Instn. Elec. Engrs.*, April 1963, pp. 157–9

15. A general study of the world's future water needs is R. FURON, *The Problem of Water*, London 1967

Bibliography

ADAMS, R. M. (1) 'Agriculture and Urban Life in Early Southwestern Iran' in *Science*, Vol. 136, 13 April 1962, pp. 109–22
(2) *Land Behind Baghdad*, Chicago 1965
ADDISON, H. *Land, Water and Food*, London 1961
AGUILA, A. DEL. (1) 'Unas Presas Antiguas Españolas de Contrafuertes' in *Las Ciencias*, 1949, pp. 185–202
(2) 'The Teaching of Professional Hydraulics in Spain' in *Revista de Obras Publicas*, VII Congreso Internacional de Grandes Presas, 1961, pp. 6–8
ALZOLA Y MINONDO, P. *Las Obras Publicas en España*, Bilbao 1899
ANDREOSSY, A. F. *Histoire du Canal du Midi*, Paris 1804
ARGUELLES, DON J. C. *Diccionario de Hacienda*, Madrid 1833
ASHBY, SIR T. *The Aqueducts of Rome*, Oxford 1935
AYMARD, M. *Irrigations du Midi de l'Espagne*, Paris 1864
BAIRD-SMITH, R. *Italian Irrigation*, 2 vols, plus volume of maps and plans, London 1855
BECCERIL, E. 'Heightening of existing dams' in *Transactions*, Sixth International Congress on Large Dams, 1958, Vol. 1, pp. 387–427
BELLET, H. *Barrages en Maçonnerie et Murs de Réservoirs*, Paris 1907
BENOIT, F. 'Le Barrage et l'Aqueduc Romain de Saint-Rémy-de-Provence' in *Revue des Études Anciennes*, Tome XXXVII, Bordeaux 1935, pp. 331–9
BOSTON SOCIETY OF CIVIL ENGINEERS. *Contributions to Soil Mechanics, 1925–1940*, Boston 1940
BOWEN, R. L. and ALBRIGHT, F. P. *Archaeological Discoveries in South Arabia*, American Foundation for the Study of Man, Baltimore 1958
BROSSÉ, L. 'La Digue du Lac de Homs' in *Revue d'art oriental et d'archéologie*, Institut français d'archéologie de Beyrouth, Paris, Vol. 4, 1923, p. 234
BRUNHES, J. *L'Irrigation dans la Péninsule Ibérique et dans l'Afrique du Nord*, Paris 1902
CASADO, C. F. 'Las Presas Romanas en España' in *Revista de Obras Publicas*, VII Congreso Internacional de Grandes Presas, 1961, pp. 357–63
CASTRO GIL, J. DE. 'El Pantano de Proserpina' in *Revista de Obras Publicas*, 1933, pp. 449–54
CAVANILLES, A. J. *Observaciones Sobre la Historia Natural, Geografia, Agricultura, Poblacion y Frutos del Reyno de Valencia*, 2 vols, Madrid 1795, 1797
CELESTINO GÓMEZ, R. 'Cronología de las Fábricas no Romanas del Pantano de Proserpina' in *Revista de Obras Publicas*, 1943, pp. 558–61

CENTRAL BOARD OF IRRIGATION AND POWER. *Development of Irrigation in India*, New Delhi 1965

COMISION INTERNACIONAL DE RIEGOS Y DRENAJES. 'Notice of Cuarto Congreso' in *Revista de Obras Publicas*, 1959, p. 402, Photo. 5

CONDER, C. (1) 'Kadesh' in *Palestine Exploration Fund Quarterly Statement*, 1881, p. 163 *et seq.*
(2) *Syrian Stone Lore*, London 1896

CONDIT, C. (1) *American Building Art; the Nineteenth Century*, New York 1960
(2) *American Building Art; the Twentieth Century*, New York 1961

COYNE, A. (1) 'Les Barrages—Differents Types et Mode de Construction' in *Le Génie Civil 1880–1930*, 50th Anniversary Issue, November 1930, pp. 202–5
(2) *Leçons sur les Grands Barrages*, École Nationale des Ponts et Chaussées, Paris 1943
(3) 'Arch Dams: Their Philosophy' in *Journal of the Power Division*, ASCE, Vol. 82, 1956, Paper 959
(4) 'New Dam Techniques' in *Proc. Instn. Civ. Engrs.*, Vol. 14, 1959, pp. 275–90

CULLEN, A. H. *Rivers in Harness; the Story of Dams*, Philadelphia 1962

CURZON, G. N. (1) 'Leaves from a Diary on the Karun River' in *Fortnightly Review*, London, May 1890, pp. 488–95, 696–715
(2) *Persia and the Persian Question*, 2 vols, London 1892

DELOCRE, M. 'Mémoire sur la forme du profil à adopter pour les grands barrages en maçonnerie des réservoirs' in *Annales des Ponts et Chaussées, Mémoires et Documents*, 1866, 2ᵉ semestre, pp. 212–72

DE SAZILLY, M. 'Sur un type de profil d'égale résistance proposé pour les murs des réservoirs d'eau' in *Annales des Ponts et Chaussées*, 1853, Vol 2, pp. 191–222

DUSSAUD, R. 'La Digue du Lac de Homs et le "Mur Égyptien" de Strabon' in *Monuments et Mémoires*, Vol. 25, 1921–2, pp. 133–41

ESTRADA, F. DE. *Reseña historica sobre las aguas con que se riega la huerta de Alicante*, Alicante 1860

FERRENDIER, M. 'Les Anciennes Utilisations de l'Eau' in *La Houille Blanche*, 1948 (Part 1, pp. 325–34; Part 2, pp. 497–508), 1949 (Part 3, pp. 121–33), 1950 (Part 4, pp. 769–87)

FORBES, R. J. (1) *Studies in Ancient Technology*, 9 vols, Leiden 1955–64
(2) *Man the Maker*, London 1958

FROIDOUR, L. DE. *Lettre à Monsieur Barillon, contenant la relation et la description des travaux qui se font en Languedoc pour la communication des deux mers*, Toulouse 1671

GEOGRAPHICAL HANDBOOK SERIES. *No. 524, Iraq and the Persian Gulf*, Naval Intelligence Division, London 1944

GIOVANNONI, G. *L'Architettura nei Monasteri Sublacensi*, Rome 1904

GLICK, T. F. 'Levels and Levellers: Surveying Irrigation Canals in Medieval Valencia' in *Technology and Culture*, 1968, Vol. 9, No. 2, pp. 165–80

GLUECK, N. (1) *The Other Side of the Jordan*, New Haven 1940
(2) *Rivers in the Desert*, New York 1959

GOBLOT, H. (1) 'Le rôle de l'Iran dans les techniques de l'eau' in *Techniques et Sciences Municipales*, February 1961, pp. 39–52

(2) 'Sur quelques barrages anciens et la genèse des barrages-voûtes' in *Revue d'Histoire des Sciences*, Tome XX, No. 2, April–June 1967, pp. 109–40

(3) 'Kébar en Iran sans doute le plus ancien des barrages-voûtes: l'an 1300 environ' in *Arts et Manufactures*, No. 154, 1965, pp. 43–9

GOMEZ-PEREZ, F. 'Mexican Irrigation in the Sixteenth Century' in *Civil Engineering*, Vol. 12, No. 1, 1942, pp. 24–7

GRAEFF, M. 'Rapport sur la forme et le mode de construction du barrage du gouffre d'Enfer, sur le Furens, et des grands barrages en général' in *Annales des Ponts et Chaussées, Mémoires et Documents*, 1866, 2ᵉ semestre, pp. 184–211

GUENOT, M. 'La Stabilité des Digues du Réservoir et du Contre-Réservoir de Grosbois alimentant le Canal de Bourgogne' in *Annales des Ponts et Chaussées*, Tome 4, 1949, Paper No. 25, pp. 467–91

HADFIELD, C. (1) *The Canals of Southern England*, London 1955

(2) *The Canals of the East Midlands*, Newton Abbot 1966

HARRINGTON, E. L. 'Early Irrigation Structures in San Antonio Area' in *Civil Engineering*, Vol. 18, No. 6, June 1948, p. 380

HATHAWAY, G. A. (1) 'Dams—their effect on some ancient civilisations' in *Civil Engineering*, Vol. 28, January 1958, pp. 26–31

(2) 'How Dams Serve Man's Vital Needs' in *Transactions*, ASCE, Vol. CT, 1953, pp. 476–88

HELLSTRÖM, B. 'The Oldest Dam in the World' in *La Houille Blanche*, May–June 1952, pp. 423–5

HINDS, J. (1) '200-Year-Old Masonry Dams in Use in Mexico' in *Engineering News-Record*, 1 September 1932, pp. 251–3

(2) 'Continuous Development of Dams since 1850' in *Transactions*, ASCE, Vol. CT, 1953, pp. 489–519

HOUTUM-SCHINDLER, A. 'Note on the Kur River in Fars, its Sources and Dams, and the Districts it irrigates' in *Proc. Royal Geographic Soc.*, Vol. 13, 1891, pp. 287–91

IBARRA Y RUIZ, P. *Estudio Acerca de la Institucion del Riego de Elche y origen de sus aguas*, Madrid 1914

IBN-AL-BALKHI. *Description of the Province of Fars*, Royal Asiatic Society Monographs, XIV, London 1912

ICOLD (International Commission on Large Dams). *World Register of Dams*, 4 vols, Paris 1964

IMAMUDDIN, S. M. *Some Aspects of the Socio-Economic and Cultural History of Muslim Spain*, Leiden 1965

IONIDES, M. G. *The Régime of the Rivers Tigris and Euphrates*, London 1937

JACOBSEN, T. and ADAMS, R. M. 'Salt and Silt in Ancient Mesopotamian Agriculture' in *Science*, Vol. 128, 21 November 1958, pp. 1251–8

JUSTIN, J. B. and TALEGHANI, K. 'Iran develops its rivers' in *Civil Engineering*, Vol. 25, March 1955, pp. 153–7

JUSTIN, J. D. *Earth Dam Projects*, New York 1932

KEATING, R. *Nubian Twilight*, London 1962

KEDAR, Y. 'Water and Soil from the Desert: Some Ancient Agricultural Achievements in the Central Negev' in *The Geographical Journal*, Vol. 123, 1957, pp. 179–87

KENNARD, J. 'Sanitary Engineering: Water Supply' in *A History of Technology*, edited by C. SINGER *et al.*, Vol. IV, Oxford 1958, pp. 489–503

KLEMM, F. *A History of Western Technology*, London 1959

KOSOK, P. 'The Role of Irrigation in Ancient Peru' in *Proc. 8th American Scientific Congress*, Vol. 2, Washington 1942, pp. 169–79

LANSER, O. 'Die Anfänge des österreichischen Talsperrenbaues' in *Blätter für Technikgeschichte*, Vienna 1960, pp. 150–71

LAZARO URRA, J. 'La Presa de la Albuera de Feria' in *Revista de Obras Publicas*, 1936, pp. 218–19

LE GÉNIE CIVIL. 'La Rupture du Barrage de l'Habra, près de Perrégaux' in Tome XCII, 1928, No. 11, pp. 256–9 ; No. 12, pp. 283–5

LEGGET, R. F. 'The Jones Falls Dam on the Rideau Canal, Ontario, Canada' in *Transactions of the Newcomen Society*, Vol. XXXI, 1957–9, pp. 205–18

LELIAVSKY, S. (1) *Irrigation and Hydraulic Design*, Vol. III, London 1960
   (2) *Uplift in Gravity Dams*, London 1958

LE STRANGE, G. (1) *The Lands of the Eastern Caliphate*, London 1905
   (2) *Baghdad during the Abbasid Caliphate*, Oxford 1900

LLAURADO, A. *Tratado de Aguas y Riegos*, Madrid 1878

LOWE-MACONNELL, R. H. *Man-Made Lakes*, London 1966

LYNCH, H. B. 'Notes on the present state of the Karun River' in *Proc. Royal Geographic Soc.*, 1891, Vol 13, pp. 592–5

MARKHAM, SIR C. R. *Report on the Irrigation of Eastern Spain*, London 1867

MINISTERO DEI LAVORI PUBBLICI. *Grande Dighe Italiani*, Rome 1961

MOORE, F. G. 'Three Canal Projects, Roman and Byzantine' in *American Journal of Archaeology*, Vol. LIV, No. 2, April 1950, pp. 97–111

MUSSO Y FONTES, D. J. *Historia de los riegos de Lorca*, 1847

NOETZLI, F. A. 'Pontalto and Madruzza Arch Dams' in *Western Construction News and Highways Builder*, August 1932, pp. 451–2

NORRIE, C. M. *Bridging the Years*, London 1956

O'CONNOR, R. *Johnstown—the day the dam broke*, London 1957

OUTLAND, C. F. *Man-made disaster*, Glendale, Calif. 1963

PARKER, J. H. *The Archaeology of Rome*, Part VIII, 'The Aqueducts', London 1876

PARSONS, W. B. *Engineers and Engineering in the Renaissance*, Baltimore 1939 (reprint 1967)

PASSA, J. DE. *Voyage en Espagne dans les années 1816, 1817, 1818, 1819*, Paris 1823

PEEL, H. M. 'A Bishop's Brain-Child' in *Country Life*, 29 September 1966, pp. 750–2

PIPPARD, A. J. S. 'The Functions of Engineering Research in the University' in the *Jour. Instn. Civ. Engrs.*, Vol. 33, 1949–50, pp. 265–86

PRELINI, C. 'Some Dams of the Ancients' in *Engineering News-Record*, Vol. 87, 6 October 1921, pp. 556–7

PROCOPIUS. *The Buildings*, Loeb Classical Library, Vol. VII of the Works of Procopius, London 1961

RAISTRICK, A. and JENNINGS, B. *A History of Lead Mining in the Pennines*, London 1965

RANKINE, W. J. M. 'Report on the Design and Construction of Masonry Dams' in *The Engineer*, Vol. 33, 1872, January 5, pp. 1–2

RETI, L. 'The Codex of Juanelo Turriano (1500–1585)' in *Technology and Culture*, 1967, Vol. 8, No. 1, pp. 53–66

REVISTA DE OBRAS PUBLICAS. (1) 'El Pantano de Tibi (Presa Actual)', Año 1900, pp. 342–5, 350–1
(2) 1854, p. 155

ROUSE, H. and INCE, S. *History of Hydraulics*, New York 1963

RUDOLPH, W. E. 'The Lakes of Potosí' in *The Geographical Review*, Vol. 26, No. 4, October 1936, pp. 529–54

SALADIN, H. 'Description des Antiquités de la Régence de Tunis' in *Archives des Missions Scientifiques et Littéraires*, 3e série, tome XIII, Paris 1886, pp. 162–4

SCHNITTER, N. J. 'A Short History of Dam Building' in *Water Power*, April 1967, pp. 142–8

SCHUYLER, J. D. *Reservoirs for Irrigation, Water-Power and Domestic Water-Supply*, New York 1901

SCOTT-MONCRIEFF, C. C. *Irrigation in Southern Europe*, London 1868

SKEMPTON, A. W. (1) 'Canals and River Navigations before 1750' in *A History of Technology*, edited by C. SINGER et al., Vol III, Oxford 1957, pp. 438–70
(2) 'Alexandre Collin (1808–1890), Pioneer in Soil Mechanics' in *Transactions of the Newcomen Society*, Vol. XXV, pp. 91–104

SMEATON, J. *Reports of the late John Smeaton, F.R.S, made on various occasions in the course of his employment as a Civil Engineer*, 2 vols, 2nd edn, London 1837

SMITH, W. E. 'Byzantine Aqueduct Still in Use' in *Civil Engineering*, Vol. 1, November 1931, pp. 1249–54

SQUIER, E. G. *Peru: Incidents of Travel and Exploration in the Land of the Incas*, London 1877

STEPHENS, J. L. *Incidents of Travel in Yucatan*, 2 vols, London 1843

STEPHENS, W. B. 'The Exeter Lighter Canal, 1566–1698' in *The Journal of Transport History*, Vol. III, No. 1, May 1957, pp. 1–11

STRABO. *The Geography of Strabo*, Loeb Classical Library, 8 vols, London 1923–30

TERZAGHI, K. 'Origins and Functions of Soil Mechanics' in *Transactions*, ASCE, Vol. CT, 1953, pp. 666–96

THOMPSON, R. C. and HUTCHINSON, R. W. 'The Excavations on the Temple of Nabû at Nineveh' in *Archaeologia*, Vol. 79, 1929, pp. 103–48

TWITCHELL, K. S. *Saudi Arabia*, Princeton 1958

UHL, W. F. 'One Hundred Years of Water Power' in *Civil Engineering*, Vol. 22, No. 9, September 1952, pp. 731–5

VINCK, F. R. 'Alte Talsperren im Jemen' in *Wasser und Boden*, Vol. 10, October 1962, pp. 354–6

VITA-FINZI, C. 'Roman Dams in Tripolitania' in *Antiquity*, Vol. XXXV, 1961, pp. 14–20, Plates II–IV

WALTERS, R. C. S. (1) *Dam Geology*, London 1962
(2) *The Nations's Water Supply*, London 1936

WARREN, W. M. and RUBIN, N. *Dams in Africa*, London 1968

WEGMANN, E. *The Design and Construction of Dams*, New York 1907

WILLAN, T. S. *River Navigation in England, 1600–1750*, Oxford 1936 (reprint 1964).

WILLCOCKS, SIR W. (1) 'The Garden of Eden and its Restoration' in *The Geographical Journal*, Vol. XL, August 1912, pp. 129–45
(2) *The Restoration of the Ancient Irrigation Works on the Tigris*, Cairo 1903
(3) *The Irrigation of Mesopotamia*, London 1911

WILSON, P. N. (1) 'Origins of Water Power' in *Water Power*, August 1952, pp. 308–13
(2) 'Water Power and the Industrial Revolution' in *Water Power*, August 1954, pp. 309–16

WITTFOGEL, K. A. *Oriental Despotism*, New Haven 1957

WOODBURY, R. B. 'Indian Adaptations to Arid Environments' in *Aridity and Man*, edited by C. Hodge, American Association for the Advancement of Science, Washington 1963, pp. 55–85

WULFF, H. E. *The Traditional Crafts of Persia*, Cambridge, Mass., 1966

ZIEGLER, P. *Der Talsperrenbau*, Berlin 1911

Glossary

Abbasid. A dynasty of 37 caliphs who were the titular rulers of the Islamic empire from A.D. 750 to 1258. The Abbasid line claimed descent from Muhammad's uncle, al-Abbas.

Acre-foot. The unit of reservoir volume commonly used in Great Britain and North America. It is a volume of water whose surface area is 1 acre and whose depth is 1 foot. 1 acre-foot is equivalent to 43,560 cubic feet, or 272,000 gallons, or 1,235 cubic metres.

Adobe. A Spanish word for unbaked, sun-dried brick and a term widely used in Spanish-America.

Aguada. The name for a Mayan water-supply reservoir. *Aguadas* were built in natural depressions and lined with masonry.

Alabaster. A dense and hard form of gypsum which is white or pink in colour.

Aqueduct. Strictly speaking, this is any artificial water channel. The term is often used, however, to describe what is really an aqueduct bridge.

Aquifer. An underground water-bearing stratum which can often be used as a source of water if tapped by a well or qanat.

Arch dam. A dam curved in plan and dependent on arch action for its strength; arch dams are thin structures and require less material than any other type.

Arched dam. Essentially a gravity dam which in plan view is curved. The name 'arch-gravity' dam is an alternative.

Astrolabe. An Arabic star calculator fitted on its reverse side with an alidade and graduated circle with which to measure celestial and terrestrial angles.

Azud. The Spanish word for a low river dam. *Azud* is taken directly from the Arabic word *as-sadd*—a dam.

Band. A Persian word for dam. The word appears in India in the form bund.

Barrage. The French word for a dam or weir but commonly used in English for a large diversion dam, especially on rivers in Egypt and India.

Basalt. A fine-grained, black and hard igneous rock.

Buttress dam. A special type in which a series of cantilevers, slabs, arches or domes form the water face of the dam and are supported on their air faces by a line of buttresses.

Calzada. The Spanish word for a causeway.

Caravanserai. A roadside inn or resting place in the East used to give shelter to caravans at night.

Cenote. A Mayan water-well lined with masonry. Sometimes called a pozo.

Chain-of-pots. A water-raising device in which an endless chain or rope carrying a series of pots or buckets is wrapped round a drum at the top and dips into a river or well at the bottom. The drum is rotated by animal- or man-power and water is lifted in the pots to the height of the drum.

Chinampas. The special type of cultivated plots built on the edge of the lakes around Mexico City. See Chapter 6, Note 7.

Chultun. A Mayan underground cistern lined with masonry and equipped with one or more narrow openings in the roof through which to draw water.

Cistern. An artificial reservoir used for drinking-water. Frequently they were built underground to keep the water clean and to prevent excessive evaporation.

Contour-canal. A navigable waterway built at one level and therefore requiring no locks.

Corbelled arch. A form of arch made up from a series of horizontally placed stones which come closer together near the top of the arch and finally meet at the apex.

Core-wall. The central watertight wall of an earth or rock dam. Originally puddled clay was the material of construction; nowadays concrete is used.

Crown. The crown of an arch dam is the section furthest upstream and is therefore usually the centre of the dam.

Cusec. The standard abbreviation for cubic feet per second.

Cut-off trench. A deep excavation under the full length of a dam filled with an impervious material—puddled clay or concrete—designed to prevent or inhibit seepage under the structure.

Dead water. The volume of water in a reservoir which is below the level of the lowest outlet and can therefore never be drawn off.

Diffuser (or Draft Tube). The discharge pipe from a water-turbine to the tail-race. When running full of water it decreases the pressure at the turbine outlet and increases the machine's efficiency.

Diversion dam. A dam built across a river to divert water into a canal. It raises the level of a river but does not provide any storage volume.

Draft tube. *See* Diffuser.

Dragoman. An interpreter in any Near or Middle Eastern country where Arabic, Turkish or Persian is spoken.

Ducat. A gold coin which was first struck in the twelfth century. Subsequently it was widely used in most European countries and minted in Holland, Hungary, Germany and Spain. It was worth about nine shillings.

Dyke. A word widely used in northern Europe for two different things. It can mean a ditch or channel used for drainage *or* a wall or bank used for flood protection.

Earth dam. A massive earthen bank with sloping faces and made watertight, or nearly so, with a core-wall and usually an impervious water face membrane.

Ell. An old-fashioned unit of length with many equivalents. The English ell is equal to 45 inches.

Fascines. A long bundle of sticks, tightly bound, and used in flood-control works, river-bank reinforcement and temporary dams.

Flash-lock or Stanch. A device used to make a river navigable. A gate across the river impounds water on one side. When the gate is opened boats either 'shoot the rapids' or are hauled through them.

Flashy. A term applied to rivers and streams whose day-to-day flow is unpredictable and liable to rise and fall suddenly.

Fulling. A cloth-working process which cleanses and thickens the material.

Glacial trough. A valley, sometimes very large, cut by the movement of a glacier.

Gravity dam. A straight dam of masonry or concrete which resists the applied water load by means of its weight.

Grist-mill. A mill for grinding corn into flour.

Grits. A general term for any sedimentary rock that looks or feels gritty.

Gypsum. Hydrous calcium sulphate.

Horizontal water-wheel. A water-wheel in which the wheel is mounted in a horizontal position on a vertical shaft; also known as a Greek or Norse mill.

Hydraulic lime. A lime which will set hard in the absence of atmospheric carbon dioxide and is therefore valuable for submarine construction. Some hydraulic limes can be made from naturally occurring materials such as lias limestone and chalk marl. Otherwise non-hydraulic limes can be rendered hydraulic by the addition of natural materials such as pozzolana, Santorin earth or other volcanic materials; or by the addition of artificial ingredients such as burnt shale or clay, crushed tile or certain slags.

Hydrology. The science of the properties, distribution and circulation of water.

Inverted siphon. When a pipeline is used to carry water across a valley it is often called an inverted siphon. Whereas a true siphon operates below atmospheric pressure, an inverted siphon develops pressures higher than atmospheric.

Lias clay. A thick bed of clay found in the Lower Jurassic rocks of Britain.

Lime. When limestone is burned, carbon dioxide is driven off to yield calcium oxide or lime. Lime mixed with water will set hard in the presence of atmospheric carbon dioxide.

Mercury–amalgam process. Used to extract gold and silver from their ores by making the precious metal amalgamate with mercury.

Mesozoic. The term applied to the second of the three great geologic eras. It was preceded by the Palaeozoic and followed by the Tertiary. There were three Mesozoic periods: Triassic, Jurassic and Cretaceous.

Mitre-gates. Pound-lock gates which form, when closed, a V shape whose apex is towards the higher water level. Thus the water-pressure forces the two gates to shut tightly.

Moraines. Rock material which is carried, or has been carried and then deposited, by a glacier.

Morisco. The name usually refers to Moslems who were nominally converted to Christianity after the fall of Granada in 1492.

Mudejar. Spanish Moslems retaining their own laws and religion but living under Christian rule.

Noria. A large water-wheel fitted with a series of pots or buckets around its periphery. The wheel is driven by the current and the pots lift water to the level of the top of the wheel.

Opus signinum. A Roman plaster made of lime, pozzolana and crushed brick or tile. It was used as a water-proofing for cisterns and aqueducts; the name is derived from the town of Signia, modern Segni.

Palmo. An old Spanish unit usually taken as 8¼ inches.

Penstock. A channel taking water from a reservoir to a water-wheel or, in modern times, a turbine.

Pictograph. A pictorial sign or symbol; also writing made up of pictures.

Piping. This occurs when the force exerted on the soil in or under an earth dam by water seeping through it is greater than the soil's capacity to resist; the water in short washes the soil away.

Pore-water pressure. The pressure developed in the pores of a material due to the presence of seeping water.

Porphyry. A general and loosely used term for igneous rocks which contain large single crystals in a fine-grained mass.

Pot-wheel. Essentially like a noria except that it is driven by a man or animal rather than by the current of a river.

Pound-lock. A canal lock with double gates. Vessels can be raised or lowered by filling or emptying the pound-lock with both sets of gates closed.

Powder-mill. A mill used for the manufacture of gun-powder.

Pozzolana. Strictly speaking it is the volcanic earth, found near Pozzuoli, which can be added to lime to give it hydraulic properties. Today the term 'pozzolanas' is used for all materials, whatever their source, which have this property.

Puddled clay. Unworked clay is not watertight; if it is mixed with the right amount of sand, wetted and kneaded, it is called puddled clay and will serve as an excellent watertight lining so long as it is kept wet.

Qanat. A slightly sloping gallery driven into a hill so as to strike an aquifer; water will then flow from the aquifer down the qanat. Essentially it is a horizontal well.

Quern. A hand-mill for grinding corn.

Relieving arch. An arch, usually segmental, built into a mass of masonry or brickwork to assist in carrying the weight of material above.

Rock-fill dam. An embankment formed largely of dumped rock for stability and fitted with an impervious water-face membrane and core-wall.

Saquiyah. A pot-wheel whose axle is geared through a right angle so that the machine can be driven by an animal (usually an ox or camel) walking in a circle.

Scouring gallery. A tunnel set low in a dam through which silt can be flushed by the pressure of a full reservoir.

Shaduf. A water-raising device consisting of a swinging beam on one end of which hangs a bucket and on the other end a counter-weight.

Spillway. The opening built into a dam or the side of a reservoir through which excess water is released.

Stanch. *See* Flash-lock.

Stop-logs. A simple form of sluice gate comprising a series of wooden planks or beams, one above the other, and held in grooves at each end.

Summit-level canal. A canal which crosses high ground. Such waterways are fitted with pound-locks and a summit-level water supply usually provided from a reservoir.

Swallow-holes. These are formed when water dissolves the calcium carbonate in a pocket or vein of limestone. Should this happen in the bed of a reservoir, severe leakage can result.

Tank. Any water-storing structure, on or below ground level, and ranging in size from the very small to the huge tanks of Ceylon and India, tens of miles in circumference.

Transpiration. The process by which plants release water into the atmosphere from the surfaces of their leaves.

Travertine. A porous rock, light yellow in colour, and used by Roman builders. The main source was a huge quarry near Tivoli.

Umayyad. A dynasty of fourteen caliphs who ruled the Islamic world from Damascus between A.D. 661 and 750. The Umayyads were descended from the cousin of Muhammad's grandfather. When the Abbasids destroyed the Damascus caliphate, Abd-al-Rahman I escaped to Spain to re-establish the Umayyad line in Cordova.

Vara. An old Spanish unit equivalent to 33 inches.

Vertical water-wheel. A water-wheel in which the wheel is mounted in a vertical position on a horizontal axle; also known as a Vitruvian mill.

Voussoir. One of the wedge-shaped stones which form a segmental, elliptical or parabolic arch.

Wadi. The Arabic word for a valley which becomes a watercourse in the rainy season. In Spanish the prefix 'guad' is derived from it.

Weir. A river dam used to raise the level so as to divert water into a canal or penstock.

Index

Abbas I (the Great), 71–2
Abbas II, 72, 73
Abbasid caliphate, origins of, 77; decline of, 82, 87
Abbasid engineering, general remarks on, 82, 87–8
Abeille, French engineer, 159
Ab-i-Gargar channel, 58, 81
Abu-l-Jund canal, dam supplying, 79
Abu Simbel temples, 243
Achaemenian dams, 13–14, 56
Adda, river, 150
Adheim dam, 79–80
Adheim, river, 61
Adige, river, 153
Adraa dam, 20, 76
Adschmaa dam, 20, 76
Adud-ad-Dawlah, Buwayhid ruler, 82; his dams, 82–4
Agricola, G., 157
aguada, 133
Aguascalientes, 140, 141
Aguasvivas, river, 105
Aguila, A. del, on Don P. B. Villarreal de Berriz, 118
Ahwaz dam, 59, 81–2
Aivat Bendi dam, 52
Aix-en-Provence, 181–3
Alagon dam, 104
Albuera de Feria, see Almendralejo
Alcantarilla dam, 48, 100
Alexander the Great, and dams in Mesopotamia, 14, 56
Alhama, dams on river, 104
Alhambra, dams for, 99–100
Ali Bey Deresi, river, 51
Alicante dam, 111–13, 122, 123, 197, 198
Al-Idrisi, Moslem geographer, 90, 94
Ali Pasha, viceroy of Egypt, 188
Al-Jurjaniyah dam, 86
Allt-na-Lairige dam, 233
Almadén mercury mines, 138
Almansa dam, 108–11, 198
Al-Mansur, caliph, 77

Almendralejo dam, 120–1, 144, 163
Al-Qantara dam, 62, 81
Alresford dam, 164–5
aluminium, 230
Alwen dam, 212
Alzola y Minondo, P., on Relleu dam, 114
Amadorio, river, 114, 115
American Foundation for the Study of Man, 15
Andreossy, A. F., author of Histoire du Canal du Midi, 163
Andreossy, F., French engineer, 163
Andronicus, Byzantine emperor, 53
Aniene, dams on river, 26–32
Anio, see Aniene
Anio Novus, 29, 31, 32
Ansotegui dam, 119
Antella dam, 94
Anthemios of Tralles, Byzantine engineer, 54
antiquity, rôle of dams, 22–4
Antioch dam, 55
Antonelli, C., Spanish engineer, 117
Antoninus Pius, Roman emperor, 50
Appleton dam, 217
Aqua Claudia, 31
aqueducts: near Mérida, 47; near Constantinople, 50, 53; Moslem, 87; in Valencia, 93; in Mexico, 139; in England, 171; in U.S.A., 183
Arabia: Sabaean dams, 15–20, 76; Himyarite dams, 20; Jewish dams, 20, 76; Moslem dams, 20, 76
Aranjuez, 117
arch dams: Roman, 32–6; structural behaviour, 33, 66, 222–4; Byzantine, 54; Mongol, 65–71; site suitability, 66, 154, 239; Safavid, 72; Spain, 113–15, 116; absence in Mexico, 144; N. Italy, 154–6;

France, 181–3, 239; stress analysis, 207, 210, 222–4, 237–9; Central America, 207; U.S.A., 207–9, 222–3; Australia, 209–10; trial-load method, 222–3, 238; European attitude to, 223–4; cupola dams, 224, 228; safety record, 224; first in Britain, 232; Scotland, 233; model tests, 238
arched dams: structural behaviour, 33–4, 191; Spain, 108–13, 116; N. Italy, 156; France, 199–200; value of curvature, 201, 204, 209; U.S.A., 209, 228
Ardashir I, Sassanian king, 56
Ardashir II, 59
Arezzo dams, 151
Arguis dam, see Huesca dam
Aristobulus, Greek historian, 14
Arizona, 144–5, 228
Arles, 33, 159
Arno, dams on river, 151
Arrowrock dam, 228
Ashby, Sir T., British archaeologist, 27, 28
Assiut dam, 221, 229
Assyrian dams, 9–12
Aswan dam, 221, 229, 243
Atabeg Chauli, Persian ruler, 85
Atcherley, L. W., mathematician, 219–20
Atrush river, Sennacherib's dam on, 12
Augsburg dam, 157
Austin dam, 217–18
Australian arch dams, 209–10
Austria: dams in Dolomites, 156; dams for fish-rearing, 156–7
Avignonet dam, 218
Avisio, river, 156
Aymard, M., French engineer, on Almansa dam, 110; his book, 110, 129, 203